The Software Hiring Hand

The Software Developer's Guide to Conducting a Job Interview

Michael Kahn

Offices in Canada, USA, Ireland and UK

Book sales for North America and international:
Trafford Publishing, 6E–2333 Government St.,
Victoria, BC V8T 4P4 CANADA
phone 250 383 6864 (toll-free 1 888 232 4444)
fax 250 383 6804; email to orders@trafford.com

Book sales in Europe:
Trafford Publishing (UK) Limited, 9 Park End Street, 2nd Floor
Oxford, UK OX1 1HH UNITED KINGDOM
phone +44 (0)1865 722 113 (local rate 0845 230 9601)
facsimile +44 (0)1865 722 868; info.uk@trafford.com

Order online at:
trafford.com/06-2217

10 9 8 7 6 5 4 3 2 1

ACKNOWLEDGEMENTS

I would like to thank my technical and editorial review panel for their efforts and time spent reviewing this book. In particular, I would like to thank the following people for their efforts and excellent feedback:

Len Colavito

Marc Leonetti

Ernest Moyer

Carlos Ramirez

I would also like to thank Mom and Dad for getting me my first computer and giving me the opportunity to start a career in software development.

Finally, I would like to thank my wife Roz for indulging me in yet another one of my crazy ideas.

TABLE OF CONTENTS

1 PREFACE

In any organization, people are the most important resource. The success of the organization is directly tied to the people within it. To address the need to staff companies with the best personnel possible, many books and articles have been written about interviewing techniques. Most of these works were written by Human Resources (HR) professionals or engineering managers. This book is written from a different perspective.

This book is written on a peer level to the software developer. Why is it useful to consider this perspective? As technology continues to evolve and the number of programming languages, hardware platforms, and standards continues to increase, it is becoming necessary for a hiring manager to rely on team members for evaluating candidates on a technical level. This book is designed to be something a software developer can quickly read and immediately use to improve his interview skills. Since those roles are typically undertaken by a HR professional or hiring manager, rather than the peer, topics such as salary negotiation and the writing of a job posting are not covered in this book. However, this book is intended to be a useful reference for a hiring manager as well as something a hiring manager can provide to his or her team to better equip the team for conducting interviews.

When I was asked by my supervisor to conduct interviews, I began to study the interview process. I read all the books I could locate on the topic and began to refine the interviewing process. After a few iterations, it became refined into something that yielded improved results. Of course, no interview technique is foolproof. Furthermore, since people and working conditions change over time, due to their maturity, as well as possible changes in the job description or working conditions, a poor candidate today may be an excellent candidate in a few years.

My interest in interviewing first came while in college, and like many college students, I became an interviewee. I had much of my interview experience that way, and only after being in the work force for a number of years did I get to be on the other side of the desk. Since I had been on so many interviews myself, when I first began to interview people, I naively figured that I knew how to do it. This is a common myth among technical professionals. The truth of the matter is that while having ample experience as an interviewee may be helpful, there are other considerations that deem it worthwhile to study the topic from the viewpoint of the interviewer.

As someone involved in the hiring for your company, it is in your best interest to bring in the best people for the job. This book provides some techniques and tips for doing just that. I encourage you, after reading this book to choose and modify the tips as needed to fit your hiring needs and come up with your own interviewing plan that works for you.

2 INTRODUCTION

If you are reading this book, then chances are that you are or will be participating in the process of interviewing people for software positions at your company. Congratulations! It is an honor and a most vital responsibility to which you have been assigned. You have the opportunity, the privilege, and the duty to do your best to help your company acquire good candidates. In sports, it is often said that good players help make other good players on the team play even better. In a software development environment, the same is true. Bring in the right people for the job, and everybody wins; the candidate wins, the company wins, and since you're part of that team, you win too.

Successful interviewing of software professionals, however, requires much more than simply qualifying technical competency. There are various other aspects to consider, including "soft skills" such as teamwork skills, communication skills, and organization skills, to name a few. Then there are other "qualities," such as a strong work ethic, attention to detail, common sense, quick thinking, and being proactive, just to name some.

This book provides you with effective techniques for interviewing software professionals. There are many great books on general interviewing of professionals. This book is intended to be a supplement to those books, not a replacement. While many books give important details on formatting an interview, negotiating a compensation package, and other important aspects, they don't provide specific detail for the software development industry. This book provides guidance for assessing the technical and non-technical skills needed for a variety of different software positions. An overview of the basics of interviewing is provided for background, but other general HR topics, such as negotiation of the compensation package are left to the wealth of other books written on the topic. It is intended that this book provide the interviewer with useful techniques that increase the chances of a successful hire, creating a winning situation for everyone, including you.

Organization of the Book

In Chapter 3, we talk about the attributes of a candidate. What types of things should we look for in a candidate? Depending on the role that needs to be filled, the answer differs. However, so we can determine which attributes we value the most, it is useful to get the attributes established early on. The attributes are used for designing the interview and performing the candidate evaluation.

Chapter 4 discusses the attributes of positions, what types of positions there are in software development and the kind of attributes a candidate should have to fill these positions. Because I believe that there is a parallel between sports teams and software development teams, where applicable, I like to use analogies from the sports world. Both have different positions, various roles to fill, and intense competition to try to beat. Like sports teams, software development teams function best when the team members work well together and their skills complement each other to form a strong all-around team.

In Chapter 5 modern interview theories and techniques are discussed. In some ways, interviewing a software professional is similar to interviewing other professionals, such as accountants, lawyers, and restaurant managers. This chapter provides an overview of generic techniques that are useful in the interviewing of professionals. If you desire more information on interviewing techniques in general, I also encourage you to read some other books, such as those included in the bibliography at the end of this book.

In Chapter 6 some prescreening techniques are dealt with. These fall into two categories, resume evaluation and phone screening. With the hectic schedules that most software development teams are trying to meet, it is important to be as efficient as possible in conducting interviews. Since you can eliminate some candidates without the need for an in-person interview, proper prescreening saves considerable time.

In Chapter 7 interview planning is discussed. This entails how the interview is conducted, who the candidate will talk to, and in what order. Some suggested interview plans are provided. Of course, you can customize these as you see fit for your organization.

In Chapter 8 we discuss the interview questions. I start by talking about a few things you should *not* ask during an interview. This is just as important as knowing what you should ask. Then we discuss technical questions used to gauge technical competency. The other attributes and soft skills can also be evaluated during this process. Some time is also devoted to "selling the job." After all, an interview is a two-way street. You are evaluating the candidate, and the candidate is evaluating you, so it is essential to be aware of this important aspect of the interview.

Finally, once all the candidates have completed the interview process, Chapter 9 discusses some techniques for candidate selection.

Why Formal Interviewing Techniques Are Often Not Applied

A common situation in many organizations is that managers and engineers, overloaded with their day to day work, often don't devote the necessary time to interview preparation. Being busy is a common reason. Another reason interview preparation is not done is that some people assume that they can "just tell" about a candidate, or have a strong "gut feeling" or intuition that guides them in determining if a candidate is appropriate. I'll first address the "intuition" factor. While it is true that some people may be a good judge of character, this is not enough. It is essential to have a systematic approach. As you will see, there are many qualities that need to be evaluated when considering a candidate that an overall "gut feeling" can't adequately measure.

In the methods discussed in this book we first determine which attributes of the candidate are deemed most important for the position that needs to be filled. I use the term *attributes* of a candidate to be a generic term for all aspects of a candidate, including work experience, education, and character traits. We also examine common software development roles and which *role attributes* are critical to effectively carry out the job function of those roles. Again, the term *role attributes* is a generic term for all aspects of a position, including the technical expertise, domain knowledge, and soft skills. Armed with this knowledge, we can craft an interview plan to extract the information we need and apply it accordingly. If you do this, you'll be much better off than if you just go with your "gut feelings."

I fully understand the daily pressures of a software development team struggling to meet schedules. Who has time to conduct interviews? Many times I've experienced a case where the interviewer just scans the resume as he is walking down the hall to meet the candidate, with no real idea of what he is going to ask, and still thinking about the other work related things that he needs to get done. I think a major reason this happens is that people don't fully realize what is at stake. However, when you consider the impact of a bad hiring decision, it only makes sense to take some time for proper interview planning.

The Impact of a Bad Hiring Decision

Let's take a brief look at what happens with a hiring mistake. If your company needs to replace an employee that isn't performing, there are many costs involved in replacing the employee. First, in all but the most drastic situations, it is known that the employee isn't working out for some time before he is let go. This time is often used to build a "paper trail" of evidence of poor performance so that the company has documentation to back up the firing in the event of any legal issues that should arise. For the sake of example, suppose that time is three months. This means that for three months, a dud is on the payroll, basically "stealing" his paycheck by collecting it and not contributing. This often means that some other team members are covering for that employee that isn't pulling his weight. This demoralizes and reduces the effectiveness of the other team members, lowering overall team productivity.

When the three months are up and the employee is let go, there are other costs that may come into play. Severance pay is often based on time spent with the company, and in some cases can be several months' pay or more. Oftentimes, benefits are provided for the employee for a few months as well. Taking the example of an engineer earning $85,000, getting four months' severance pay and benefits:

Four months' salary: $28,333

Four months' benefits: $6,400 (assuming $400 per week for 16 weeks)

Total Severance Costs: $34,733

So that's about $35K just to get rid of the employee. That doesn't even take into account replacement costs. One cost that always is incurred is employees' time taken away from work duties to perform interviews. Other costs that may also be incurred include recruitment advertising, employment agency fees, and candidate travel expenses. Being conservative, let's assume that you interview all local candidates and use some web-based and local newspaper advertising. Suppose you end up interviewing four candidates.

Advertising costs: $2,000

Time spent reviewing resumes and conducting four interviews:

Manager plus three engineers @ 5 hours each = 20 hours @ $50.00 per hour

= $1,000.

Total Recruiting costs: $5,000.

Note that the estimate of $5,000 for the recruiting costs is *very* conservative. If an employment agency is used, recruiting costs can go much higher.

Total replacement costs: $39,733.

So it costs us almost $40,000 to replace the employee that was brought on as the result of a bad hiring decision. But these are only the "tangible" costs. There are intangible costs as well. These include lost productivity, lost opportunities due to not meeting deadlines, and other costs that are not possible to directly measure. It can get even worse though. In many companies, the employee has to do something drastic before getting canned. So what happens to an employee with mediocre performance, but shows up every day for work? In many cases he stays on indefinitely, since the company won't fire him. In this case, the intangible damage can be huge. There have been cases where an employee clearly isn't working out, but the company will not terminate him.

If a poor hiring choice is made, the company may end up living with it indefinitely. In other words, the person is a poor performer, but not poor enough to get fired, so he remains and continues doing damage for extended periods.

This is why it is so critical to implement an effective hiring plan. This greatly increases your chances of recruiting people that are a good fit for your company and department. The information and methods presented in the following chapters help you to create an interview plan for your needs.

3 CANDIDATES

In professional sports, teams assess their needs in the off-season, and this often influences what types of positions they will draft for the next season. For example, if a football team needs to improve the passing game, they may likely acquire offensive players such as quarterbacks, receivers, and offensive linemen in the off season. In baseball, a team needing to improve pitching will recruit pitchers. Of course, there are different kinds of pitchers. There are starters, expected to last seven innings or so, and closers, expected to throw hard for an inning or two. A pitcher may be well suited for one role, and not so well suited for the other role. Therefore, when acquiring players, careful consideration is given to the role the player is expected to fill.

Note that we are discussing the candidate attributes first, before discussion the types of positions. You might wonder why I chose to start with the candidate attributes rather than talking about the attributes of the positions. Addressing that matter, imagine that your friend wants to buy a vehicle, but he lacks knowledge about the various types of vehicles that exist. You could start the discussion by asking him what type of driving he will do and then try to determine an appropriate vehicle. However, there are advantages to first introducing the various types of vehicles, such as economy cars, sport utility vehicles, and minivans, for example. By getting exposure to all types of vehicles, as well as various vehicle attributes (e.g. gasoline powered, diesel powered, hybrid, etc....), it then provides your friend with knowledge about the various possibilities available , and he can then evaluate his needs with this in mind.

In a similar manner, we will first discuss the candidate attributes. Then, in the following chapter, we discuss position attributes, and how they relate to the candidate attributes. As mentioned previously, I use the term *attributes* of a candidate to be a generic term for all aspects of a candidate, including work experience, education, and character traits (soft skills).

Experience

The experience of the potential employees can be divided into four categories:

- Time in Workforce
- Skill Relevance
- Organizational
- Leadership

Time in Workforce

The first is experience as defined by the amount of time in the workforce. This can be broken into three levels: entry level, mid level, and senior level. I consider entry level to be less than three years of work experience in the field of software development. Mid level is three to seven years, and over seven years is considered senior. Entry level candidates are often used when building a team for the future. It is to be expected that the entry level candidates likely will not have the hands-on experience, judgment, or possibly even the maturity to handle some software development roles. But they are well suited for other roles. The key to remember with recruiting entry level people is that they don't stay entry level forever. Think of entry level recruiting as an investment in the future, with the intention that the entry level people you hire today will be strong contributors in a few years. Since they don't have a lot of practical experience, your focus should be more on what kind of persons they are, their character traits, and soft skills. Academic performance should also be considered to gauge capabilities and interests. In the chapter on interview questions, we devote some consideration to entry level candidates.

A mid level candidate has been in the software development profession for three to seven years. This candidate has some hands-on experience and should know something about software development. However, as noted later, the amount of time isn't always the best factor to determine if a candidate is qualified. As compared with an entry level candidate, the hope is that with a mid level candidate, there will be a reduced learning curve.

Finally, there is the senior level candidate, with over seven years of experience. It is generally expected that these candidates will be able to make a contribution with a relatively short learning curve.

It is worthwhile to point out that a candidate with more time in the workforce does not necessarily have more maturity, skill, or competence than a candidate who has less time in the workforce. As observed later in this book, there are many technical skills and non-technical (soft) skills that need to be evaluated for a given position. Using techniques to evaluate these needed skills, you may find that sometimes a candidate that happens to have less time in the workforce is a better fit than someone with more. The important thing is to objectively evaluate the candidate on his qualifications and attributes, regardless of how many years of work experience he has. The following table summarizes time-in-workforce experience levels:

Table 3-1

EXPERIENCE LEVEL (Time in workforce)	DESCRIPTION
Entry Level	Less than three years experience. Expected to join the team and grow with the company.
Mid Level	Having three to seven years experience. Expected to join the team with a relatively short learning curve, bringing knowledge acquired from other companies.
Senior Level	Over seven years experience. Able to make a contribution with a relatively short learning curve.

The first thing you should ask yourself when considering a candidate with X years of experience is: "Experience doing what?" Is nine years of experience better than three? It depends. If you need someone for Linux driver development, then who would serve your needs a better, a candidate with three years of Linux driver development or someone with no Linux experience but nine years of COBOL experience? The type of experience we are talking about here is *Skill Relevance*. In this contrived example, the answer is easy: the person with three years would be the logical choice for first consideration. However, it is possible to be misled by skill relevance, so we must take a careful look at the different types of skill relevance and what they mean.

Skill Relevance

The skill relevance of the potential employees can be divided into four levels:

- General
- Functional
- Application (Domain)
- Specialist

A candidate with general relevance has experience that pertains to software engineering or development in general, but not in the same area as the job opening. For example, the job opening may be for a graphical user interface (GUI) programmer, but the candidate presents experience in embedded programming but no GUI experience. The job opening may be for Java, but the candidate has mostly C experience. This is the lowest level of skill relevance.   However, it is important to note that you do not necessarily want to rule out candidates who only have an experience level of general relevance.

In the next chapter we study the attributes of roles (job positions) in detail.   It is seen that in many cases general relevance is perfectly acceptable. In many cases a sharp, talented person with general relevance experience joins a team and within a few months is outperforming some of the more seasoned team members. This is because having the ability to learn fast is a very valuable "soft skill" that may outweigh the lack of directly related experience.   After all, if one is are proficient in a particular programming language or scripting language, it is likely that he will be able to learn another in relatively short order. So while it's not appropriate for every position that needs to be filled, there are many cases where general relevance suffices.

A candidate with functional relevance has experience that pertains to similar implementation techniques but not necessarily in the same industry. For example, the job description may require RTOS (real time operating system) embedded programming for an avionics application, yet the candidate has RTOS embedded programming experience for multimedia applications. Functional relevance pertains to familiarity with the development techniques, rather than the application that uses those techniques.   The distinction is critical. Functional experience can be very valuable to a team. This is especially the case if other members of the team have extensive application (domain) experience. In this case the addition of members with strong functional experience complements the team, and it is likely that in time the new member will acquire some application experience.

A candidate with application relevance has experience that pertains to a similar application as yours. This is different than the functional relevance described previously. Just because a candidate has high application relevance doesn't mean he has high functional relevance. For example, suppose the job opening is for multimedia software development, more specifically for UI (user interface) design, and the candidate has considerable television development experience, but with embedded software development for tuner control algorithms and no experience with UI. An important point to realize about application relevance is that in many cases it isn't that relevant at all. There is the story of an interviewer who overvalued a candidate that had considerable application experience but was deficient in other areas. In effect, the interviewer was "smitten" with the fact that the candidate had experience in the same industry and recommended him for hire. The candidate was lacking in other areas, such as functional experience. The candidate was hired, and in fact it turned out to be bad decision. The candidate was not equipped to handle the role he was expected to fill, which required considerable functional experience. There are some cases where application experience is critical, but in many cases it isn't. So while it is always valuable to have some application experience, it is important that we do not overvalue it when making hiring choices.

A candidate with specialist relevance has experience that pertains to specific algorithmic knowledge. For example, if a job position requires advanced knowledge in a field such as data compression or data encryption, then not every software developer is going to be able to fit this position. You'll need someone with high specialist relevance. Note that it is possible to have specialist relevance and not have much functional or application relevance.

Sometimes this is fine. For example, suppose you had a job position that required a data encryption expert to develop crypto algorithms in C for an embedded processor for multimedia applications. You find a candidate with twelve years of crypto development. The candidate worked exclusively in military applications (almost zero application relevance), and only developed in Ada on special workstations used by the military (low on functional relevance). However, if other members on your team can provide the needed application and functional relevance, then this candidate may make a great addition to your team. The candidate's strength (crypto knowledge) may complement the strong C skills and multimedia application experience of other team members to help improve your team overall. Of course, not every position requires specialist relevance, so in some cases it won't matter. Furthermore, a specialist may not be well suited for roles outside his specialty. What happens when you need to ask the specialist to do some other work? Keep this question in mind when considering a specialist. If at some point you would want your crypto expert to contribute in other areas when there is no active crypto crisis, you will want to consider the broader capabilities of the specialist candidate. The following table summarizes skill relevance experience levels:

Table 3-1

EXPERIENCE LEVEL (skill relevance)	DESCRIPTION

General Relevance	Experience pertains to software engineering or development in general, but not in the same area as the job opening.
Functional Relevance	Experience pertains to similar implementation techniques, but not necessarily in the same industry.
Application (Domain) Relevance	Experience pertains to a similar application as yours.
Specialist Relevance	The job description requires detailed algorithmic knowledge of a particular area (e.g. data compression, and the candidate has considerable knowledge of data compression algorithms).

Organizational Experience

The organizational experience of a potential employee can be divided into these categories:

- Lone Wolf

- Team Internal

- Multi-Discipline

- External Customer

- External Vendor

The Lone Wolf, as the name implies, is a software developer that primarily works alone with accountability only to his or her supervisor. As software complexity increases, software development has become a predominantly team sport. Today, teamwork skills are essential in many software development positions. Nevertheless, there are positions that don't require large amounts of team interaction. These may include R&D positions or maintenance and support roles in smaller or mature projects. Another role to which a Lone Wolf may be well suited is that of an isolated subsystem on a project. For example, in an embedded system, the bulk of the software may be written in C on a general purpose processor. Yet, there may also be one or more special purpose processors or DSPs executing their own microcode. Often times, one person is responsible for the DSP microcode. This person needs to be able to work independently, yet also is part of a larger team.

If your position does require teamwork interaction on various levels and your job candidate has primarily Lone Wolf organizational experience, then you need to be aware of this. You should not necessarily disqualify a candidate just because of this. However, you will want to ask questions to gauge how he might react in a team environment. We discuss these questions in an upcoming chapter.

Someone with Team Internal organizational experience has worked in a software development team, but primarily interacted only with other members of that team. This is quite a step up from the Lone Wolf, as he likely had to deal with a number of team issues, such as determining if a problem was in his code or someone else's, dealing with configuration management and version control systems with multiple developers, and assisting other team members as needed. For many software developer positions, Team Internal organizational experience will suffice.

For other positions, Multi-discipline organizational experience is preferred. In these positions, a software developer will often deal with other departments within the company. This may include other software development teams, other engineering teams, such as hardware (electronic) or mechanical engineering teams, quality assurance teams, manufacturing, marketing and sales, just to name a few. Working with multiple departments requires certain skills that aren't as critical in a Team Internal or Lone Wolf situation. For example, various groups often are competing for resources and may have their own agendas internal to that group. A multi-discipline team member should have some situational awareness, and political savvy.

The last two, External Customer and External Vendor, are similar. As the names imply, the former position deals with external customers to the company, and the latter deals with vendors. Note that there is often overlap among Multi-discipline, External Customer, and External Vendor roles. A project manager or team leader will often spend considerable time in all three aforementioned roles. I divide external contact into the External

Customer and External Vendor roles because there are a few key differences.

Someone with External Customer experience has spent time dealing with *customers*, clients that have a business relationship where they receive goods or services from the company in exchange for money. Someone with External Customer experience has a feel for customer service, making the customer feel like he is being attended to. He also has a sense for what not to say to a customer, what information is confidential, and maintains a sense of professionalism at all times when dealing with external customers.

Someone with External Vendor experience has spent time dealing with vendors. I use the term *vendor* in the generic sense of a company that provides goods or services to your company in exchange for money. In software development, a common vendor interaction is with the technical support team of another company. If your team will be relying on support from a vendor for a chip or a software package, then you may want to consider the External Vendor experience of your candidate. Dealing with the technical support (or field application engineers (FAEs)) of a vendor company requires care. On one hand, you are their customer, and it is in their interest to help you...to a point. Often, a software development team will use the vendor's software as a starting point to integrate the vendor's software and/or hardware into the product under development. As your application begins to deviate from the supplied reference software, your FAE may reply that the problem doesn't occur in the supplied reference code, so it must be something you are doing wrong. What will you do? The project is on the line, and the vendor isn't jumping through hoops to help you. While there is no single correct answer that applies to all scenarios, an employee with solid External Vendor experience will have some sort of plan. For example, he should try to collect as much data as possible to provide to the FAE to give him the best opportunity to find a solution. The employee may try to reproduce the problem in the reference platform if possible or provide a way to allow the FAE to duplicate the problem using the application. On an ongoing basis, the ideal employee would try to develop rapport with the FAEs and treat them with respect. Where possible, the worker would try to resolve issues directly with the FAE. Only when there is a dispute about a delivery or unresolved issue would the employee escalate the issue and get management involved. This reduces the risk of animosity at the engineering level.

In some projects, you might be dealing with multiple vendors and relying on their support to complete your project. If that applies to you, you will definitely want to consider the amount of External Vendor experience your candidates have. The following table summarizes organizational experience levels:

Table 3-2

EXPERIENCE LEVEL (organizational)	DESCRIPTION
Lone Wolf	Worked primarily alone, with accountability only to supervisor.

Team Internal	Worked as part of a team, but only interfaced with members of his group.
Multi-discipline	Worked as part of an engineering team and also interfaced with other teams, such as SQA, Manufacturing, Field Service, Marketing, or Sales.
External Customer	Interfaced with external customers
External Vendor	Interfaced with external vendors

Leadership Experience

The leadership experience of a potential employee can be divided into these categories:

- None
- Trainer
- Mentor

- Team Leader

- Manager

A candidate with "none" has never worked as a leader in any capacity. This is to be expected with candidates having zero to three years of experience. It sometimes happens that someone with many years of experience has never worked as a leader. This isn't necessarily bad. However, if someone with many years of experience has never worked as a leader in any capacity, it could indicate that

a) He does not desire a leadership position, or

b) He was never given an opportunity for such a position, but would consider it, or

c) He is incapable of taking on such a position.

Other considerations aside, if you have a senior level candidate with no leadership experience, you would want to probe to find out what the reason is. If your position requires some leadership, or you are looking for some potential future leaders in your organization, then you may want to avoid candidates with a reason of "a" or "c." In some positions, leadership is not a big part of the job, and then it may not matter much if the candidate doesn't have any leadership experience.

A candidate with Trainer leadership experience has helped show new colleagues how to perform a task or use a system. This is short-term help that stops once the trained colleague can understand the task. While this isn't day to day leadership, it still requires effective communication skills, patience, and a working knowledge of the job. A candidate with Trainer experience may be on the way to a higher level of leadership.

The next level of leadership characterized is Mentor. This differs from Trainer leadership in that, while Trainer leadership involves the training of specific tasks or procedures, Mentor leadership provides a higher level of guidance. In addition to providing specific instructions, a Mentor provides information that allows more junior members to help themselves. Instead of telling them the answers, they are told where they might look to find the answers. Another aspect of the Mentor may involve some motivational discussions with more junior members to help keep them enthusiastic and develop their skills. Mentoring positions are often informal, in that nothing may exist in writing that says "Jane Smith, you will mentor John Doe as part of your job duties." More often, it is an informal and implied relationship between senior and junior members. Because it is informal, people often don't including mentoring experience on their resumes, so it may be necessary to ask questions during the interview to determine if the candidate has Mentor level experience. You might ask, "Have you mentored any more junior members? If so, what was your approach?" There are no definitive correct or incorrect answers. However, possible responses may include motivation, helping the junior employee understand his role in the department and the role of the department in the company, teaching approaches to solving problems rather than solving problems for him.

Mentoring is an essential part of any team that has a mixture of junior and senior members. I am grateful to those who have mentored me early in my career, and as I became more senior, I have made an effort to mentor junior members. Because mentoring helps the productivity of the junior members, thereby improving the performance of the entire team, and since mentoring requires many skills needed in other, more formal types of leadership, it is worthwhile to consider Mentor experience in candidates who are expected to fill a formal leadership position in the future.

The next leadership level is that of Team Leader. The Team Leader has the responsibility of giving and overseeing assignments for a group of people. He doesn't have administrative authority over the team. Therefore the Team Leader doesn't get involved in discussions about vacation days, salary, and other administrative issues. As he interacts with a variety of other groups, such as a quality assurance group, marketing, or manufacturing, the Team Leader typically requires Multi-discipline organizational experience.

The Manager is responsible for a group of people in an administrative capacity. He has to deal with the administrative issues of the staff. Of course, it is possible that the Manager also acts in a Team Lead capacity, having technical jurisdiction, as well as managerial jurisdiction over the team.

It is less common to find someone with Management leadership experience applying for software development jobs, but it does happen. The two main reasons are as follows:

REASON A:

Candidate wants to get back into software development.

REASON B:

Candidate lost a management job and is now considering all available options in his job search.

Why you might consider a "REASON A" candidate:

Management isn't for everyone, so if a candidate realizes it isn't for him and wants to go back to software development, there is a reasonable chance he will stay the course. This is good if you're looking for a software developer long term.

Why you might consider a "REASON B" candidate:

If you have a "fast track" position, where you are looking for a leader in the near future, a "Reason B" candidate may be a good fit. It often happens that a person with management experience in one industry when transitioning to a new industry has to gain some experience "in the trenches" as a software developer for one to three years and then may have the opportunity to move into a leadership position.

An important point about the candidates who have most recently been managers and haven't done much technical work lately is that their software skills may be a little rusty. As is noted later in the book, this affects the interviewing strategy. The following table summarizes leadership experience levels:

Table 3-3

Experience level (Leadership)	DESCRIPTION
None	Never worked as a leader in any capacity.
Trainer	Helped show new colleagues how to perform a task or use a system. This is short-term help that stops once the colleague can understand the task.
Mentor	Helped more junior colleagues with assignments on an ongoing basis.
Team leader	Responsible for technical assignments of a group of people.
Manager	Responsible for a group of people in an administrative and possibly technical capacity.

In addition to work experience, there is also educational experience. In a similar manner, educational experience may have direct relevance, such as a degree in Computer Science, or a more generic relevance, such as a Science degree (e.g.: Physics, Chemistry, etc….), or an Arts degree, (e.g.: English, Music, Math, etc….), or no degree at all. The educational requirements and policies determining them vary among jobs and companies. In some applications, a technical educational background is essential. In other cases, it is not. If a candidate has strong work experience, then the political science degree he earned 12 years ago shouldn't necessarily disqualify him from the opportunity to interview. I have encountered a number of talented software development professionals that did not have a technical degree. Remember, just as it is important to screen out people that won't make a good fit, it is also important not to falsely screen out someone that could make a good fit. In the end, you need to consider the specific position and the candidate seeking it, to determine if his educational experience meets your criteria. In summary, if the position will allow it, consider being open to candidates with degrees outside of computer science or engineering, especially if the work experience of the candidate makes him a potentially good fit for your organization.

Soft Skills

In addition to the classifications of experience mentioned above, there are also other skills we should consider. These skills are not specific to software development, but are important in almost any position. These skills are less tangible and focus more on the specific qualities of the candidate. They are often referred to as "soft skills."

Communication Skills

Communication Skills are important in almost every facet of life, so it should be no surprise that they are also important in a software development environment. Ineffective verbal and written communication slows things down and can cause mistakes, making ineffective communication quite costly to a software development project. Ideally, we want all our candidates to have excellent communication skills. As we see in upcoming examples, some positions demand more communication skills than others.

Learning Skills

The ability to quickly learn new things is a very desirable skill for a candidate. In your particular hiring situation, you must have some expectation of how long it will take for the candidate to come up to speed on the job duties. Obviously, you want to select a candidate that is able to learn quickly enough, such that the candidate's learning curve matches your expectations.

Diligence

Doing quality work is important. Will your candidate make the necessary effort to make sure the job is done correctly? Does he take pride in his work?

Proactive

While having a worker that has good communication skills, good learning skills, and diligence is nice, it is even better if, in addition to the soft skills just mentioned, he is also proactive. This means he may look outside of the scope of his work and identify problems or opportunities that impact the department or company. It means the phrase "that's not my job" doesn't come up too often in his vocabulary.

Attitude

An overall positive attitude is helpful in solving problems, taking on new tasks, and working with others. In many cases, you can glean something about the candidate's attitude by asking him why he wants to leave his current job, or what he doesn't like about his current job.

Discretion

This soft skill is particularly important when communicating outside the company. If the candidate will be dealing with customers, a general sense of what should be said and what should *not* be said in front of customers is important. Note that I am not suggesting that we want a dishonest candidate. I am simply stating that discretion, the ability to know how to handle sensitive information among various groups, is a valuable soft skill, especially in the case of dealing with people external to the company, such as customers and vendors.

Because soft skills are often less tangible than technical skills, they are more difficult to evaluate than technical skills. For example, you may get a good feel for how well a candidate knows COBOL, but how can you get a strong sense of how much discretion he possesses? There is no single technique to probe the soft skills.

However, sample questions are provided, along with the soft skills they are intended to examine. These can be used as a starting point for building your own interviewing plan that meets your needs for a particular hiring situation, the main goal of this book.

In this chapter, we defined the various types of experience and soft skills that software developers may have. We are now ready to do similar definitions for some of the typical software development positions. Once we have all our definitions in place, we have the tools for discussing which candidates may be good for our positions, providing means for a system for screening resumes, the first step in the candidate selection process.

4 POSITIONS

In the previous chapter, we discussed attributes that define the candidate. Now we examine various roles within a software development environment and see what attributes may be most important for them. Note that some of these roles may overlap.

Entry Level Software Developer

The Entry Level Software Developer is presumably filled by a candidate with Entry Level experience, meaning zero to three years experience. Since he has a limited amount of experience, there is not much of a work history to go on. However, hiring at the entry level is often necessary to grow a team. In a sense, hiring entry level people is an investment in your company's future. Just as a sports team drafts rookies in hopes that they learn the system and contribute more and more, so that in a few seasons they are key players, a software development department needs to bring in entry level talent to grow the department and also to allow others to advance to new positions.

Usually, entry level developers start out doing basic tasks under fairly close supervision. Although they lack experience, there are various attributes, including "soft skills" that are desirable in the entry level candidate. Your interview of the entry level candidate should be designed to evaluate these attributes.

These attributes include:

- Eager to learn

- Taking responsibility for his actions

- Good interpersonal skills

- Enthusiastic about the subject matter

With the entry level candidate, there is usually a higher weight placed on the academic record than with someone who has been in the workforce for several years. Here are some of the factors to consider with entry level candidates:

- Grade Point Average (GPA)

- Extracurricular activities

- School Projects

- Any work or co-op experience

GPA isn't everything, but if a candidate has a high GPA, then chances are he:

- Has a good propensity for learning

- Has the discipline required to get good grades

- Is interested in the subject matter of his major

What about candidates that do not have a stellar GPA? There are a few other ways to analyze it. For example, you can consider GPA in the major. If the GPA in the major is acceptable, then it is still worthwhile to find out why the courses outside the major posed a problem for the candidate. On one hand, it may be understandable that a candidate devoted most of his time and energy to courses in his major, placing those courses at a higher priority than non-major courses. On the other hand, even non-major courses are important and should be taken seriously. If the candidate doesn't take a course seriously because he doesn't think it is relevant to his major, then does that mean that he won't put much effort into a small assignment you give him on the job? In the workplace, it is important to do even small tasks well. Ideally, we want an entry level candidate with that attitude. If you gave an entry level candidate an assignment to log some data for a program under test, you wouldn't want him to do a sloppy job because he felt it wasn't a critical piece of the program. In fact, in software development every task should be handled with care and professionalism. So if the GPA in the major is much higher than that outside the major, it is worthwhile to ask why.

Another way to look at GPA is throughout the college career. If the data is available to you, you may want to see if the candidate's grades dramatically improved in his junior and senior years. If a candidate had a 2.9 (out of 4.0) GPA overall, but the GPA of the last two years was a 3.4, it shows that the candidate made some improvements during his college years, and, rather than just the overall 2.9 GPA, you would also want to take that into consideration.

Another part of academic life is extracurricular activities. A candidate that achieved a high GPA while managing a few extracurricular activities such as participating in a club, sport, or part time job, indicates that the candidate can manage multiple tasks simultaneously.

In addition to GPA and activities, some entry level candidates graduate with relevant work experience through a college work-study program. This is helpful for you as the interviewer as it provides more material for asking questions that can help you judge their abilities.

The key point is that when considering someone for an entry level software position, in light of the fact that there is little work experience, we are generally more interested in soft skills. Basically we want to know:

- Can he learn?

- Will he fit here?

- Is he capable of contributing in time?

In the chapter on interview questions, a section is focused on an interview plan for entry level candidates. This plan will help you extract the information to form a judgment on those questions.

Maintainer

The Maintainer works with primarily existing code in a mature product. The work entails mostly fixing bugs and minor feature enhancements. Maintenance programming is often looked down upon and thought of as an unfavorable assignment. However, in many companies maintenance plays an important role. Often, the software that is in the maintenance phase is "paying the bills" for the company, being a major source of income. As mentioned throughout this book, maintenance is also an excellent way for a new software developer to learn about your company's products and way of doing things. Given these important points, I devote some additional focus on the Maintainer role by providing a few key attributes that are especially important for maintenance work.

- Attitude that maintenance programming is a great place to start

- Able to learn the workings of code that he didn't write

- Surgical Strike Capability

- Diligence in researching change history

ATTITUDE

Many software developers do not wish to make a career out of maintenance programming. Rather than the upkeep of old software, they often desire to be more involved in the development of new products and be involved in the development of new software. Positions involving new development are often thought of as more glamorous.

Then there are others who enjoy debugging and the feature add-ons that the maintenance programmer has to deal with. For the latter type, the attitude should not be much of an issue. However, for the former type, even if he desires to be on the new development team, a maintenance position is often a good place to start. Starting a new hire in a maintenance position provides him an opportunity to learn about the application and industry. This is important if the candidate is light on Application experience. Typically, the product in maintenance is a predecessor of the newer development products, so having some knowledge of the previous product can be helpful in understanding the new product. This makes maintenance a good starting task for a developer entering a company.

Therefore, you want to avoid the candidate that thinks that maintenance programming is "beneath" him. For a new employee, maintenance programming offers an excellent opportunity to learn the industry and some of the company's more mature products.

Of course, you should clearly communicate to your candidate if the position you are filling is a long-term maintenance position or a growth position with the potential to move into new opportunities such as new software development or a project leadership role. If you bring on someone with aspirations beyond maintenance programming, he will be frustrated if it appears he is stuck in maintenance land indefinitely. If he understands that he will likely be in maintenance for six months to a year with potential for a new role after that, then he may be more receptive to viewing the maintenance assignment as a learning opportunity rather than a prison sentence.

ABILITY TO LEARN

An important skill of the maintenance programmer is the ability to quickly learn about software that he didn't write. There are some developers that have a great command of the software they have written but don't like going in unfamiliar neighborhoods. For a new hire being brought on for a maintenance project, the whole thing is an unfamiliar neighborhood. Your maintenance programmer should have a system for learning unfamiliar code. This may include reviewing whatever documentation is available, taking his own notes, reviewing the code, stepping through the code in a debugger, making minor changes to the code to confirm his understanding, adding console prints to the code to observe operation, or some other technique. It doesn't really matter what the technique is, so long as he has some sort of conscious approach to learning unfamiliar code that works for him.

SURGICAL STRIKE

A maintenance programmer is typically working on a mature product that is often already deployed. Therefore, the maintenance programmer must be prudent in making changes. The candidate should be sensitive to the risk of the change and attempt to code the change in a manner that does not cause unnecessary risk. Sometimes, a change that is appropriate during the development cycle is not appropriate during the maintenance phase. The successful maintenance programmer recognizes that. There are some that feel that every sub optimal piece of code should be rewritten to fix every bug. While the code may not be great, sometimes it is just not appropriate to rewrite big chunks of code, as, given the current phase of the project, the risk of doing that may be inappropriate. Of course, each situation is different, with its own set of circumstances that require evaluation. However, given the maintenance state of the product, the maintenance programmer should be aware of the scope of the changes, the risk of the changes, and select the software modifications that have an appropriate level of risk. After all, if you have a leaky roof, yes, sometimes you rip off the old roof and replace it with a new one. In other cases, however, it is more appropriate to apply a patch to your existing roof. The circumstances, such as severity of the leak, how long you plan to live in the home, repair budget, and other considerations, come into play. The successful maintenance programmer will understand that the same applies in software development and be aware of the situation that pertains to the maintenance project.

DILIGENCE

As the maintenance programmer makes changes to fix bugs, he will sometimes come across a code segment that is causing the problem, and if that code segment were to be removed, the problem would be fixed. Where possible, the maintenance programmer should try to determine who added that code, why, and when. This will give insight into the appropriate fix. Sometimes, a cut-and-paste error may cause extra lines of code to appear where they shouldn't. In other cases, that code was put in place to fix a particular problem and shouldn't be removed without knowledge of that problem. Therefore, if it is necessary to remove or modify those lines of code, the old problem can be retested and addressed another way.

As it can usually show when the lines were added, and by whom, the version control system may provide many of these answers. Sometimes a comment entered when the file was committed via the version control system will refer to the problem the code change was intended to fix. Sometimes, the maintenance programmer will be able to walk to the next cubicle and ask the person who made the original change. In other cases, that person is long gone, and nobody at the company remembers anything about why it happened. In that case, the maintenance programmer has to do his best to make a proper change. The important point is that he should always perform his "due diligence" and try to determine why changes were added, whether those changes are the result of a bug. This applies to any phase of software development, but since he won't be familiar with the history of the code, in a maintenance project it is especially important for a new hire to be aware of this.

Implementer

The implementer carries out fairly detailed instructions of more senior members. Tasks may include implementing software from a design document, designing and performing unit tests, and participating in the integration of his module into the system.

In addition to the technical skills required, teamwork skills are essential in this role. The software developer should have awareness of team development, using a version control system, and be sensitive and courteous in regards to "breaking" the build. That is to say, if a software developer commits code that doesn't compile, or compiles but does not function, it could halt the progress of an entire team. The Implementer (and anyone else who commits code into the repository) should make every attempt to keep the build working. This includes being careful about his code commits and taking precautions such as using a "diff" utility to confirm the changes on each file to be committed, verifying the compile and build with a clean checkout, and performing a reasonable amount of unit and system testing before committing the code. For an Implementer role, a candidate with functional relevance skill experience is important. Since he tends to follow instructions from more senior members who presumably have the application experience, the Implementer doesn't necessarily need a lot of application relevance experience.

Architect

The architect role involves the design of software modules, API calls, inter-module communication, and data structures. The position requires a high level of functional relevance experience. Some application relevance experience is also desirable. Since he will likely be communicating his designs to teams of implementers, good written and verbal communication skills are important. The architect may also be an implementer. However, in many cases, at least a portion of his design is delegated to other team members to do the implementation. Architects need to see a bigger picture than an implementer. In addition to being knowledgeable in software design principles, they need to consider various factors such as code maintainability and testability.

Systems Engineer

The Systems Engineer is typically removed from the actual writing of software. He is involved in higher level details such as system requirements. In larger projects, it is very common to have a team of systems engineers dedicated to developing system requirements, whereas smaller projects may not have such a group. Systems Engineers are especially important when the developed product is part of a large system. As he needs to know "what this thing is supposed to do," the systems engineer needs application relevance experience. A Systems Engineer should have an awareness of higher level issues that sometimes extend beyond the actual writing of code. Such issues may include upgradeability, security, and compatibility with the other parts of the system. As he spends a good deal of time reading and writing specifications, excellent written and verbal communication skills are valuable in this role.

Software Support/Field Application Engineer

The software support role requires interaction with customers. Application relevance is important here. The in-house software support role requires troubleshooting customer issues, mostly over the phone or via Internet communication. If your position requires some in-house technical support, the candidate should have a systematic approach to determining the circumstances of a failure in the field. You might ask a question such as, "How would you respond if a customer called you and simply said, 'My system isn't working'"? Responses should include follow-up questions to the customer for more information. Possible responses may include:

- What kind of system do you have?

- When did it stop working (ascertain if the system was ever working)?

- What was going on at the time it stopped working?

The Field Application Engineer may do a mix of in-house and on-site support. Regardless of in-house or on-site, the successful software support candidate needs to have a "customer service mindset." As he is representing the company, he has to handle himself professionally. The software support role obviously benefits from someone with External Customer organizational experience.

SDK Developer

The SDK developer develops Software Development Kits (SDKs) that are used by other groups or external customers. This role requires functional relevance, application relevance, and excellent written communication skills. Typically, the SDK developer has extensive experience as an implementer and possibly some experience as an architect as well.

UI Developer

The User Interface (UI) developer creates user interfaces. The good candidate will have a working knowledge of various human factors such as ease of use and designing UIs to reduce operator error. The UI developer ideally has application experience. In most products, the UI is one of the most heavily scrutinized and criticized parts of the software. The VP of marketing most likely won't be studying implementation of the bubble-sort algorithm, but he may use the product and comment on the UI. Therefore, you may want to consider someone who can receive feedback on his work without taking it too personally or becoming defensive. This is a good quality in general, but it is critical on highly visible areas such as the User Interface.

Embedded Developer

The embedded developer works on a special purpose system, such as a mobile phone, music player, or digital camera, just to name a few. Typically, embedded software programs are somewhat tied to the hardware they run on. Therefore, the embedded developer may work closely with hardware engineers, so some multi-discipline experience may be helpful. Embedded developers need good debugging skills. This is because, in many cases, a standard integrated debugger is not available. Sometimes there isn't even a console for outputting print statements. Therefore, some debugging creativity is desirable.

Special Ops

It is rare that you will actually see a classified advertisement for a "Special Ops" software position, but they are more common than you might think. These are the developers that are called upon to do demos, prototyping, proof-of-concepts, and special builds for a particular customer. These developers have resourcefulness and creativity and can think on the fly to "get something working tomorrow." In many cases, the code they write may not conform to the company coding standard, and the people that are good at "special ops" don't always fit in with the mainstream software developers. Yet many companies have a few "special ops" people on their software development teams. In their most effective use, they generate prototypes that demonstrate proof of concept, and then their work is turned over to a different group to create a more formal design and implementation. The special ops candidate has very high functional experience and may have high application experience too. Special Ops positions often do not require significant organizational experience, and the successful Special Ops candidate may even be a Lone Wolf in some cases.

In some cases, the Special Ops candidate has Specialist experience. For example, if you have a need for a crypto expert, his crypto expertise is critical. Functional experience and Application experience may be secondary. You probably have others on your team that have this knowledge and can fill in the blanks where the crypto expert is lacking.

Integrator

The integrator is responsible for bringing various pieces of software together and making sure the components operate properly. Often, an integrator has to work with third-party libraries for which he does not have the source code. This creates unique challenges for the integrator. For example, he may have to prove that a problem exists in a particular third-party library. For this reason, External Vendor experience is often desirable in an integrator. In general, as well as high functional experience, he needs good teamwork and interpersonal skills.

Quality Assurance / Test

The Quality Assurance (QA) role involves developing test plans and procedures, executing test plans, and reproducing problems and finding the pattern that causes them. He often communicates with software development engineers to discuss their findings. He requires excellent observational skills. So he can attempt to reproduce problems that they encounter, he needs to be aware of the circumstances and conditions during testing. If he isn't observation oriented, he will miss important details that will make the problems he finds hard to reproduce, making them take longer to fix.

Ideally, the experienced QA candidate will have some application experience and good teamwork and communication skills.

Temporary Employees

It is worthwhile to note that in many ways selecting a temporary employee (contractor) is similar to selecting a permanent employee. You still want to focus on needed attributes. However, the focus of a temporary employee is typically short term. That is, you are likely to focus more on the exact technical skills you need. Just to illustrate the point, if you have a project involving embedded Linux and you are hiring a contractor to help with some of the detailed Linux issues, you will likely expect quick contributions and a short learning curve. To do that, the candidate will likely need prior experience with Linux. With a permanent employee, you need to consider the hire as more of an investment in the future. While relevant experience is still important when selecting a temporary employee, such other factors as soft skills and leadership potential, important in selecting a permanent employee, are not as scrutinized.

Remote Employees

In this age of globalization, it is becoming increasingly common to have remote employees. These may constitute some "Lone Wolves" that are working out of their home, or a small group of people in a remote office. In other cases, it may be a large remote office, perhaps separated by thousands of miles, several time zones, and an ocean or two.

The remote Lone Wolf may work out of his home, or a small office. The qualifications are overall similar to that of a Lone Wolf that is working in your office. However, the abilities to work independently and responsibly take on a greater importance.

It is becoming more common for organizations to have remote software development offices. From an upper management point of view, there is a desire to take advantage of a well educated and highly skilled work force that is abundant, and cost effective for scaling an organization. The remote software development team may be collaborating with your team on the same code base, or they may be autonomous, working on a distinct area or project.

A remote software development team may provide a way for an organization to remain competitive, and handle multiple projects that would not be possible otherwise. In order to be successful, the remote office must be staffed with the right people. For the most part, a given position requires the same attributes, regardless of if the work is performed in your office, or the remote office. A remote office, once established, will often handle its own staffing. However, when starting up a remote office, the hiring of the senior software developers requires careful attention. In theory, there would be no difference in the interview approach.

However, in reality, there are some constraints that may make it difficult to conduct interviews in a traditional manner. Namely, due to distance, a face-to-face (F2F) meeting with candidates may be prohibitively expensive and time consuming. Furthermore, due to time zones, it may be difficult to even conduct telephone interviews. If you are staffing a large remote office, it may simply not be possible to have all the candidates interviewed by people in your office. In many cases the remote staffing is handled on site, via a local recruiter. However, your group should qualify the senior software developers that will be starting the remote office. The senior software developers will likely be overseeing the work of the more junior members, and they will likely be doing extensive communication with your office, at least to get started. It is essential that these senior software developers have the ability to oversee and mentor the less experienced developers. Otherwise, the work may go unsupervised and might not get done properly. In this case, the activity of the remote team may actually impede your project, rather than help it. Therefore, it is worthwhile for the "local" office developers to conduct interviews for the "key" senior development positions of the remote office. If F2F meetings are not possible, try conducting telephone interviews, even if you have to do it at an odd hour due to the difference in time zones. If telephone interviews are not possible, consider an exchange via e-mail, web conferencing, "virtual whiteboards", or Internet messaging (IM), since these are often the methods you are most likely to use during routine communication.

There are various recruiting agencies that specialize in staffing remote software development teams. These agencies can be very helpful in handling much of the logistics required to start a remote office. However, keep in mind that these agencies, while they may desire to bring in good people, may have other influences, such as financial gain from completing the staffing. These agencies usually do not have a stake in the particular projects of your company. Therefore, for key positions (e.g. Team Leads), it is important that your group perform some technical assessment, as well as evaluate the communication skills. The communication skills take on a new level of importance with a remote software development team. Collaboration with a remote team is extremely difficult if the communication is not good. If the senior developers of the remote team do not have strong technical and communication skills, or are not capable of mentoring and assisting the junior members, the chances of a successful collaboration with your group are greatly reduced. In summary, if you want good results from your remote team, you need to staff it just as diligently as you would a local team.

The table below summarizes the roles just discussed:

Table 4-1

ROLE	DESCRIPTION
Entry level.	Performing some of the more basic and mundane tasks, with the hope that the exposure will facilitate learning that will allow the candidate to take on more advanced assignments.
Maintainer	Someone who works with primarily existing code in a mature product. Work entails mostly fixing bugs and adding minor feature enhancements.

Implementer	Carries out fairly detailed instructions of more senior members. Tasks may include implementing software from a design document.
Architect	Designs software modules, API calls, inter-module communication, and data structures.
Systems Engineer	Analyzing requirements, designing functional building blocks.
In-house Tech Support (Software support engineer)	Responsible for troubleshooting customer issues, mostly over the phone.
Field Application Engineer	Responsible for troubleshooting customer issues, with considerable "face time" at customer sites.

SDK developer	Developing SDKs to be used by other groups or companies. May require more attention to detail and writing skills.
UI Developer	Develops User Interfaces. Considers human factors such as ease of use.
Embedded Developer	Develops software for embedded devices.
Special Ops	Works on special projects, proof of concepts, demos. Needs to get something working quick, doesn't have to be pretty. Willing to cut corners, but might not write award winning code. But you don't need that for prototyping
Integrator	Bringing various pieces of software together, and inserting the glue.

Quality Assurance / Test	Responsible for quality assurance, developing test plans and procedures, reproducing problems and finding the pattern that causes them.

What does it all mean?

Note that this is a generalization of some of the more common roles in software development. Your company may have other roles, or various positions may embody multiple roles listed here. That is to say, you might have a need for a special ops integrator architect. You don't usually see that in a job description for a classified advertisement. More likely, you advertise your position for a Senior Software Engineer, or some other title. However, it is worthwhile for you to give consideration for some of the more likely roles that your candidate will need to fill in the next year or so.

Now that we have defined the attributes of candidates and their roles, we have the vocabulary to discuss which candidate attributes are critical in given roles. This will guide us through our prescreening, the subject of the next chapter.

5 PRESCREENING

The last two chapters discussed the key attributes of candidates, defining typical software positions in terms of those candidates. The table below summarizes those roles, and the critical attributes that are absolutely essential for the candidate to have.

Table 5-1

ROLE	CRITICAL ATTRIBUTES
Entry level	Academic Performance, Communication Skills, Attitude, Work Ethic, Discipline
Maintainer	Functional Experience, Application Experience
Implementer	Functional Experience, Teamwork skills
Architect	Functional Experience, Application Experience, Teamwork Skills, Communication Skills

Systems Engineer	Application Experience, Teamwork Skills, Communication Skills
In-House Tech support (Software support engineer)	Application Experience, Customer Skills, Communication Skills
Field Application Engineer	Same as for In-House
SDK Developer	Functional Experience, Application Experience, Communication Skills
UI Developer	Application Experience, Human factors skills, Ability to take feedback without feeling hurt or defensive
Embedded Developer	May require multi-discipline teamwork skills and excellent debugging skills.
Special Ops	Functional experience or Specialist experience

Integrator	Functional Experience, Application Experience, Communication Skills, Teamwork Skills, Customer Skills.
Quality Assurance / Test	Application Experience, Communication Skills, Observational Skills

Some explanation is in order here. First, the positions that you need to fill may not directly map to the roles listed above. They may be combinations of multiple rules. For example, you may need someone who is an Integrator but is also Special Ops, or an Architect that is also a UI developer.

For each role, there are listed skills that are critical for success in that role. This means that without possessing all those skills in reasonable amounts, their ability to perform in that role will be limited. In many cases, interviewers underestimate the importance of "soft skills" in software developers. For example, a technical wizard who has poor teamwork skills or poor communication skills won't do well in an Integrator role. However, the same person might be OK in a Special Ops position. Many interviewers focus on technical knowledge without giving sufficient attention to the Soft Skills. However, it is both possible and worthwhile to attempt to assess the level of these Soft Skills for the roles that require them.

Note that just because an attribute is not listed does not mean it is not desirable. For example, it should not be inferred from the previous table that someone in a Maintainer role does not need to have any communication skills whatsoever. There are many desirable traits that all employees should have, including communication skills. The critical attributes are meant to be the subset of all attributes that are *most* important to be successful in that role. Quite a few effective Maintainers don't have great communication skills. It is wonderful when they have strong communication skills, but even when they don't, they may still be successful in the Maintainer role.

Similarly, Attitude is a critical attribute of the Entry level position. Obviously, attitude is important in every position. However, because the entry level candidate is light on work experience and all that the interviewer has to go on is academic performance and the soft skills listed, it takes on a higher importance in an Entry Level role.

I can hear some of you out there saying, "No way man! The Maintainer has to have teamwork experience, and the Application experience isn't that critical, because the senior members of the team can help out in that area when needed."

And to that I say, "Excellent!" You know the needs, the team dynamics, and corporate culture of your company and department. This table is intended as a starting point. You should customize this table, removing and adding roles and attributes as appropriate for your hiring situation. Once you have edited the table as you see fit, you will use it as a prescreening aid for evaluating resumes of potential candidates.

Prescreening the Resumes

Time is a very valuable asset to any software department. The whole point of prescreening is to save time by screening out people that in all likelihood will not be a good match for the positions you are trying to fill. Keep in mind that one of the quickest and most effective ways that you can help your team is by helping to bring in the right people. An in-person interview will easily take up several man-hours, especially if several team members will be interviewing the candidate. Therefore, thorough and systematic telephone screening is essential. I recommend a two stage screening process:

1) Resume Screening

2) Telephone Screening

Using this two-stage process to qualify your candidates will serve to minimize the chance of bringing in an unqualified candidate. After posting your job advertisements, you likely have a bunch of resumes to look through. Using a methodical approach, you want to get that large pile of resumes narrowed to a smaller pile that you will move to the next phase, the telephone screen.

The first step is to review the job description, and consult your modified version of table 5-1 to remind yourself of the critical attributes of the candidate for the available position. Next, do a cursory scan of the resumes (and cover letters), removing from consideration those that clearly don't meet the requirements. For example, if you've determined you want someone with at least seven years' experience, don't consider the resumes with two years. It seems that regardless of how many years you list on the job requirements, you always get some that clearly don't meet them. Note that "remove from consideration" doesn't necessarily mean the candidate is not a good match for another position within your organization. Some resumes may be kept on file for other positions.

Don't consider candidates that submit poorly written resumes and those with an abundance of grammatical or typographical errors. Attention to detail is important. If they are sloppy with their resume, it could be an indication of how they will be with other things they have to write. Don't consider excessively long resumes. My absolute limit is three pages. If a candidate can't provide a *summary* of experience in three pages or less, then he likely has some communication issues. I recommend that you do your resume reviews in a few short sessions, rather than one marathon. Reading resumes is fatiguing. When you are fatigued, your judgment isn't good. Instead of reading resumes for six hours straight on Monday, read them for an hour in the morning and an hour after lunch for a few days, so that by Thursday you've completed your first pass through the list.

Once you have completed the first pass, review the remaining resumes again, this time more carefully. In this pass, read the resume in its entirety, highlighting items that would be good discussion areas with the candidate. Write on the margin, and also generate some questions on extra pieces of paper if necessary.

Remind yourself of the critical attributes of the position. Consider your needs in terms of functional experience, application experience, organizational experience, and whatever other attributes are critical. Eliminate any other resumes that upon further examination are lacking in a critical area. The remaining resumes should now meet the minimum criteria for your position. If you're unsure if a candidate meets the minimum requirements, err on the side of caution and don't place it in your rejection pile just yet. Use the telephone interview to confirm the candidate's attributes. Based on the outcome of that telephone interview, you can determine if he gets discarded or moved on to the following round of the playoffs, an in-person interview.

Depending on how many resumes are left in your pile, it may be worthwhile to give them one more scan to rank them. You can rank them in ascending order or make an "A" list and a "B" list. Rank them by which is the best fit, considering not just critical attributes, but all attributes that might be desirable. For example, Application experience might not be considered critical for an implementer role, but if a candidate has some, that would be a plus that would advance that candidate towards the top of the list. Look at the accomplishments listed on the resume and see how they map to the soft skills of interest. For example, do any of the accomplishments suggest being proactive, diligent, creative, etc...?

Once you have your list ranked or sorted into an A-list and B-list, you are ready for the next phase, the telephone screen.

Telephone Screening

The purpose of the telephone screen is not to make a decision about hiring the candidate or to judge every aspect of his soft skills. This is quite difficult to do over the telephone. The sole purpose of the telephone screen is to narrow your list even further, so that the only candidates that visit your company are those that have a reasonable chance of being a good fit.

The telephone interview should be arranged at a time convenient for both parties. It is often convenient to use e-mail to set up the details of the telephone call including the time, who will initiate the phone call, and what number to use.

For currently working candidates, doing a telephone interview during business hours may be challenging. It is often difficult to find privacy in the workplace for talking to a potential future employer. Therefore, try to be flexible in scheduling the interview time. If necessary, consider options such as before or after work or during lunchtime. A candidate will be more responsive if he has the privacy needed to discuss a potential job change. By being flexible with the time, you help ensure this privacy. Once the time is agreed upon, to save any telephone charges it is good etiquette for the prospective employer to offer to call the candidate. However, give the candidate the option of calling in if he prefers to do so.

As mentioned previously, the telephone interview is not meant to probe every detail of the candidate's history, but to determine if the candidate has the potential to be a good fit. Have the candidate's resume in front of you during the call, and think of a few questions ahead of time. Here are some possible discussion points:

- Reasons the candidate has left (or wants to leave) his current and previous positions
- The type of experience the candidate acquired at his current and previous positions
- Likes and dislikes about the positions he has held
- What he is looking for in his next job
- Desired salary range (if appropriate)

The telephone interview typically starts with the interviewer providing an introduction and a brief description of the current position. Describe the position, but don't reveal the critical attributes. If the critical attributes are revealed, it may allow the candidate to tailor his responses accordingly, basically telling you "what you want to hear." It is better to hear the candidate's honest sentiments during the interview, than to receive answers that match what the candidate thinks you want to hear.

Good way to phrase the description:

"We are looking for a senior level firmware engineer to participate in board bring-up, network driver development, flash driver development, and integration of third-party libraries as needed."

Bad way to phrase the description:

"We are looking for a dedicated, hard working, creative problem solver, and team player to develop software for an embedded application."

The "good way" gives an idea of the type of duties that need to be performed but doesn't reveal the specific qualities you are looking for. To the extent possible, allow the candidate to provide answers without being biased by your questions.

With the "bad way" phrasing, you've given the interviewee some traits to focus on, and he will likely tailor his responses to highlight these traits. To get the most objective answers, try not to reveal too much about the critical attributes. For example, if you need a database expert, then you will likely mention that in your discussion. If you need a database expert with excellent teamwork skills, don't mention the part about the teamwork skills in your question. Instead, let your questions about past experiences, conflict resolutions, and likes and dislikes guide you in forming an opinion about his teamwork skills. If one of his "dislikes" was working in a group, obviously this candidate is not right for your position. Even if he has excellent database skills, if the candidate won't work with the team, it does little good.

You can ask questions about anything on the resume, but try not to get too technical on the telephone. For example, you might ask: "I see you have experience writing Linux kernel patches. Tell me about that." This prompts the candidate to discuss it as appropriate over the telephone. He may discuss, at a high level, the types of things he did in this regard. However, asking too detailed a question over the telephone is a disservice to the both of you. For example, you wouldn't want to ask, "Please explain the inter-process communication between all the tasks in the current system you are working on." Questions of this type are best answered in person, most likely with the aid of a whiteboard. It is difficult to explain this level of technical detail over the phone, so save those questions for those that make an in-person appearance.

As you conduct the telephone interview, take notice of the following:

Does the candidate have sufficient verbal communication skills for the position?

Since you have done your homework, and already determined the level of communication skills required for the position, you should be able to assess this during the course of the telephone interview. If the candidate has inadequate verbal communication skills, he won't be effective on the job. Think about how often one person needs to explain something to another person on a software development team. It happens all the time. Now imagine that, because of poor communication skills, instead of taking five minutes to explain each concept, it takes five times as long...25 minutes. Then, since more time is spent re-explaining and correcting misunderstandings resulting from poor communication, the entire software development cycle is slowed down. Different roles require different levels of verbal communication. Be sure that the candidate meets the requirements for the position, and don't accept a candidate that falls below your minimum requirements.

Does the candidate understand how the position contributes to the overall goals of the department and company? In explaining his current (or previous) job, he should be able to explain how his job fits in to the grand scheme of things within his department and company. In other words, does he know why he is doing what he does? The candidate who has no awareness of anything outside his own job is less likely to be concerned about the performance of the department and company. If you don't get a good sense of this as he is discussing his job, then don't be afraid to ask the question directly: "How does your position fit in with the overall goals of your department and company?"

Do the likes and dislikes of the candidate coincide with the critical attributes of the position? If the position involves maintenance, and the candidate dislikes fixing bugs in other people's code, then you can save yourself the trouble of an in-person interview with that candidate. On the other hand, if the candidate "enjoys the challenge of finding and fixing problems, regardless of where the problem lies," then that is pretty well aligned with the critical attributes of a maintenance role.

Can your company provide opportunities on par with what he is looking for in his next job? If the candidate desires to be in management within a year or two and you know that isn't likely at your company, you are well advised to be up front about it. If an employee is misled, he may get frustrated and quit. As discussed previously, employee turnover is expensive, so be up front and realistic about the opportunities that are available within your company.

Is the desired salary range in the realm of reason for your position?

Note that this question is appropriate when managers are doing the telephone screening. If a peer is doing the screening, then the salary questions are probably off limits. The peer should defer those questions to the manager. However, for the manager, it is worthwhile to know the salary range ahead of time if possible, to see how it fits with your company's preferred compensation range. Note that there are many good books that discuss compensation negotiation, some of which are listed in the appendix of this book. Since this book's focus is on qualifying software professionals, generic topics like salary discussions are omitted, and the reader is encouraged to read some of the other good books to get more information on this topic.

Is there a discrepancy between resume and the candidate's true level of experience? Does the candidate say he has five years' Java experience, but it turns out he spent the last five years doing very rudimentary things (e.g., writing "Hello World") in Java? Be wary of any extreme exaggerations. If it is too much of a stretch, disqualify the candidate. In general, you can conduct a brief "Buzzword Reality Check" over the phone by asking a general question regarding some of the "keywords" on the candidate's resume. For example, if you see XML on the resume, you can ask, "I see you have XML on your resume. Please tell me what you've done with XML." You can determine if his experience involves designing and implementing an XML parser, reading about XML in a book, or somewhere in between. The object is not to get into super detailed technical discussions, but to see if the candidate has reasonable knowledge of the subject that he put on the resume. If he put the buzzword on the resume, then it is fair game to ask about it.

Does the candidate lack critical functional experience? If you absolutely need someone who has familiarity with Real Time Operating Systems (RTOS) and the candidate has none, then disqualify that candidate. Think carefully about what attributes you really *need* to have in your candidate. Beware of making unnecessary restrictions that could screen out a potentially good candidate.

When you have completed your list of questions and want to draw the telephone screen to a close, you can ask something like, "Is there anything else you want to say?" It signals to the candidate that the discussion is concluding and provides a last opportunity for any other discussion points or candidate attributes that haven't been covered by your previous questions.

By this time, you should now have reached one of two conclusions about the candidate:

1) The candidate seems to be a good match, and is definitely worth talking to in-person

2) The candidate is definitely not able to fill the position, or you aren't sure

In the first case, schedule an in-person interview to move this candidate to the next round in the selection process.

In the second case, you don't want to waste time bringing the candidate in for an interview for this position. However, it is important to be courteous and professional to the candidate. Unless the telephone screen was a complete disaster, I recommend phrasing something that leaves the door open for future contact. Often, a candidate that isn't a good match for a position you have today is a good match for a future position. Therefore, you want to leave that as a possibility. If you think the candidate might be a good fit in another position that may be available in the future, you might say, "Thanks for taking the time to speak with me today. We have several other candidates to speak with and will be making a decision shortly about who we will be contacting for in-person interviews. If we feel that there is a possibility of a good fit for this particular position or any other position, we can reach you at the telephone number you have provided."

Regardless of how the telephone call went, when you have completed each of your telephone screens, you should write down your thoughts on the candidate as soon as possible while the telephone call is still fresh in your mind. Once you have completed all your telephone screens, your list of resumes should be a smaller, more manageable group, and you can start to schedule your in-person interviews. If there is truly no further interest in the candidate for this position, it is good etiquette to let the candidate know fairly soon after the interview. Many companies do this via a post card that states that the candidate is no longer being considered for this position, but his resume will be kept on file. In this way, the candidate is not left hanging. He knows where things stand and can move on.

In the next chapter, we discuss some fundamental interview theory, which provides guidance in developing your own customized interview plan.

6 INTERVIEWING BASICS

Much of the information presented in this chapter is generic and applies to almost any profession, not just software development. There are many works written on basic interviewing theory, and the reader is encouraged to seek out those books for more information. The bibliography at the end of this book lists some excellent references for those that desire to learn more. For our purposes, it is necessary to review some basic interview principles to develop an interview plan that can be used for candidate selection. In this section, some sample questions are presented, and possible answers from the candidate are shown.

Interviewing Styles

There are two prevalent styles in interviewing software development professionals: the Behavioral style and the Situational style.

Behavioral

The Behavioral style focuses on past actions of the employee. The premise is that by asking about situations from current or previous work experience, you can gauge how the candidate will perform in the future. Behavioral questions are intended to probe three key areas:

1) Knowledge: Does the candidate have the skills needed to do the job?

2) Motivation: Knowing how to do the job is not enough. If the candidate is not motivated to do the job, it likely will not get done or will not be done well. Therefore, it is worthwhile to try to determine if the interests of the candidate are well aligned with the job. For example, if the job requires working in a group and the candidate likes working in a group, there is alignment between the candidate and the job as far as the team environment goes.

3) Actions: What kinds of actions has this candidate taken in previous situations? These provide insight into how the candidate may handle a similar situation at your company.

Common discussion points include:

- How did the candidate handle a difficult situation?
- Explain a time when something didn't work out as planned.
- What was the toughest problem you ever had to debug?

In general, the behavioral question is an open question that seeks to get information about a situation that the candidate previously encountered.

The general format of a behavioral question has an opening question followed by one or more follow-up questions.

Here is a simple template for forming behavioral questions:

Opening question:

"Tell me about a time when X happened. How did you handle it?"

"X happened" is a placeholder for a premise. The premise serves as a starting point for a discussion that is prompted by one or more follow-up questions:

Follow-up #1

Follow-up #2

Table 6-1 shows a few sample premises.

Table 6-1

OPENING QUESTION PREMISES
You had to struggle to meet a deadline
You missed a deadline
You found a serious design flaw late in the development cycle
You had to provide feedback to someone who didn't take criticism well

You had to quickly learn a software system that had little or no accompanying documentation

Once the candidate gives the initial response, you can probe further with an appropriate follow-up question. Table 6-2 lists a few examples.

Table 6-2

FOLLOW-UP QUESTION EXAMPLES
"What did you do next?"
"What was your role specifically?"
"What, in hindsight, could have been done to prevent this?"
"If you could do it over again, what would you have done differently?"

The pattern you are looking for is that the candidate was able to be resourceful and effective in the situations that were discussed. In cases where things didn't work out, you are looking for the ability to learn from the experience. A candidate who claims to have never made any mistakes or never learned anything from a situation is undesirable and often indicates an arrogance that can be disruptive to a team environment.

In general, there is a basic pattern, known as SAR (Situation, Action, Result) that fits the format of a behavioral interview.

Situation:

This is the background needed to understand the problem.

Action:

This is what the candidate did in response to the situation.

Result:

What happened as a result of the candidate's action? Was the result positive? Did the action drive the result? For example, suppose that in responding to a behavioral question the candidate states, "I immediately checked the vendor web site and found that a new firmware release was posted. I checked the list of fixed bugs and saw that it corrected the problem we were having." That is an example of an action driving a result. Suppose that the candidate states, "I had no idea, so I closed my eyes and hit the Enter key, miraculously, it worked!" While the result may have been positive ("it worked!"), the result did not come about because of good actions from the candidate. This is an example of a result not being driven by the candidate's action.

Ideally, the candidate will paint a clear picture of the situation, explain the action taken, and the result that followed, without much prompting from you. If he doesn't answer to your satisfaction in response to your opening question, use follow-up questions to gain more information about the situation, action, or result.

Situational

The Situational style focuses on hypothetical situations presented to the employee. The premise is that by asking him about how he would react in situations likely to be encountered in your workplace, you can gauge how he would perform there.

Below are some examples, and sample responses (shown in italics). Obviously, you would tailor your situational questions to have relevance to your workplace.

How would you handle it if a big release was due tomorrow, and you found a serious problem the night before?

Sample Response: I would collect as much information as possible on the problem, and immediately report it to my supervisor, so he could make an informed decision about how to handle it and devise a backup plan if necessary. Then, if possible, I would try to get the problem fixed in time for the release.

This answer shows that he has the proactive sense to inform people about potential problems as soon as possible. Most bosses don't like to be surprised about problems. The answer also shows that effort towards fixing the problem would be forthcoming.

What would you do if you were integrating a third party library, and you found a problem in it that was a showstopper for your project?

Sample Response: I always take a "guilty until proven innocent" approach with integration. I assume the problem is in our code first, but after performing tests to verify that it is in fact in the third party library, I would try to collect as much evidence as possible to show that and then present it to the tech support for the third party library.

This is a good question to ask someone in an Integrator role. You are looking for someone who isn't going to just "throw problems over the wall" to the tech support of the other company. You want someone who can appreciate the end goal of completing the project and will collect as much data as necessary or possible to convince the third party to investigate and repair their software, allowing your project to move forward. The "guilty until proven innocent" approach tells you that the candidate will investigate issues thoroughly on his side before getting the third party involved. When a software developer continuously gets the third party involved too quickly, it can have an adverse effect on the response of the third party. Much like "crying wolf," they tend to take you less seriously if it happens too often.

What would you do if you submitted a development schedule and were told it was too long?

Sample Response: I would take a second look at the schedule to see if there was any way to perform tasks in parallel or bring in the schedule in another way, say by reusing an existing module. However, I try to be careful and realistic when submitting a schedule in the first place. Assuming my second look didn't turn up anything, I would inquire about getting additional resources or dropping or delaying some requirements to work out a schedule in the appropriate timeframe.

There isn't a single correct answer to this question, but you are looking for some insight into thought process of the candidate regarding scheduling and what he would do if his schedule is challenged.

Describe your ideal job and why?

Sample Response: Ideally, I would like to have a job where I am exposed to a variety of technologies and aspects of the industry. I like being an expert in a particular part of a system, but I also want to continuously learn about other parts of the system.

This is a good question for learning about the motivation of a candidate and evaluating if the candidate's motivation is in alignment with the job description.

In general, the candidate should have a legitimate answer for the Situational questions. Responses along the lines of "I have no idea" or blank stares may be indicative that the candidate would not know how to handle these situations at your work place.

The ideal interview will have a mix of both Behavioral and Situational questions. Note that by changing the beginning phrase from "Tell me about a time when…" to "How would you handle …," you can easily convert most Behavioral questions to

Situational questions. For example, if you ask a behavioral question about how a candidate handled a particular situation, and the candidate responds that he has never encountered that situation, you can follow with a situational variant of the same question: "OK then, how *would* you handle that situation?"

Stress Interview (best avoided)

The Stress interview is a technique where the interviewer checks to see how easily the candidate gets flustered. It involves an onslaught of tricky (and sometimes insulting) questions. The interviewer then takes note of how the candidate keeps his composure. Software development jobs can be stressful at times. There are deadlines that have to be met. Software defects can have economic repercussions, and in extreme cases, lives are at risk from software defects. However, software developers typically aren't subjected to the same stress levels as a brain surgeon or an air traffic controller. In those cases, a stress interview may be appropriate. However, for a software developer role a stress interview where you try to fluster the candidate is generally inappropriate. This isn't to say that you don't ask some tough questions. However, you want to avoid the mentality of "let's stump the candidate." Rather, you want to focus your questions around the critical attributes of the position and assessing the candidate's strength in those areas.

Question Types

In addition to interviewing styles, there are a few fundamental question types that the informed interviewer should be aware of.

Closed Questions

These are questions that have a yes or no answer. Unskilled interviewers often ask too many closed questions such as, "Can you fix difficult problems under pressure?" Every candidate will likely answer yes, leaving the interviewer with no real information, and no way to distinguish one candidate another. Closed questions can be an effective way to confirm information such as, "You were a project lead at your last position?"

Open Questions

The majority of questions you will ask during the interview will be open questions. These questions cannot be satisfied with a yes or no answer. The open question requires an explanation in response. For example, "Tell me about a time you fixed a difficult problem under pressure." As opposed to the closed style above, phrasing the question this way will force the candidate to explain a particular situation, giving you material for follow-up questions: "Why did that happen?" "In hindsight, what could you have done to prevent the crisis from occurring in the first place?" This will help you gain insight into the candidate that closed questions usually don't provide.

Leading Questions

In general, you want to avoid leading questions, as they may mask the candidate's true response. Try to phrase your questions in an objective manner so that the candidate does not get clues to try to give the answer that you want to hear, rather than a truthful response. While discussing the position, interviewers sometimes ask leading questions accidentally. The interviewer might mention, "We are a small company, and people here have to wear several hats. Most of our software engineers are also quite handy with logic analyzers and oscilloscopes. Can you operate an oscilloscope?" You have telegraphed to the candidate that using signal measuring equipment is important to you, and then asked if he can use it. The candidate will now try to give an answer that will be in agreement with your needs.

In general, when asking a question, try not to indicate your preferred answer to the candidate. While the candidate is speaking, be mindful of your body language. You should not indicate agreement or disagreement. Some interviewers have a habit of nodding their head as the candidate is speaking or saying, "That's right!" indicating agreement with what the candidate has said. Make a conscious effort to avoid giving cues to your candidate that will have them shape their answers accordingly. Let the true sentiment of the candidate be heard!

Conversational Techniques

There are a few conversational techniques to keep in mind when conducting an interview. Using these techniques will allow you to make the most of the short time you have with the candidate.

Let the Candidate Talk

It is a widely accepted general rule that during an interview, you should be talking 20% of the time and listening to the candidate 80% of the time. This is known as the "80/20 rule." The interview is not the place for you to ramble on to the candidate about your experience. You are there to learn about the candidate's experience. You need to provide some introductory information and control the interview by the questions you ask. Remember, when you are talking, you aren't evaluating the candidate. By having an interview plan beforehand (which you will be able to create after you finish this book), you won't be rambling on while trying to think of a question.

Some interviewers tend to comment on the candidate's answers during the interview. This wastes valuable time, and as was mentioned in the section on Leading Questions, you don't want to telegraph your feelings about the candidate's answers. Don't indicate agreement or disagreement. If you have lost your place, and need to stall, you can ask, "Please elaborate" or "Why did that happen?" to keep the candidate talking while you get back on track with what you were going to ask. Make a conscious effort to abide by the 80/20 rule while conducting your interviews.

The Royal "We"

When discussing previous experience, a candidate may sometimes explain what "we did." "*We* implemented cold fission, *we* developed a nixie tube simulator, *we* did this and *we* did that…." If the candidate is answering most of the questions about what "we" did, try to determine specifically what he did. If necessary, ask, "What was your role?" or "What did you do specifically?"

Staying out of the Weeds

The candidate may get distracted or go on a rant about a particular situation. This is not good. The astute candidate will be in tune with the interviewer and gauge the appropriate level of response.

Nevertheless, sometimes everyone gets off course with what the interviewer is expecting, or time runs short and it is necessary to move on. Remind the candidate that you need to keep the interview moving.

One way to get a candidate out of the weeds and back on the road is with a realignment question such as, "Given our time constraints, I think it would be best if we move on to another topic. I'd now like to discuss [topic]. Would that be OK?" The candidate will likely agree, and the interview progresses.

Phrasing your Questions

Avoid long compound questions when interviewing. For example, a question such as "Tell me about the types of systems you worked on, how long it took you to learn each one, and which one you liked the most, and the least, and why." If you want to hear the candidate's response to all of these questions, it is best to administer the questions in small doses, allowing the candidate to adequately concentrate on each individual question.

Dealing with Panic

Sometimes, a candidate will panic or get a "deer in the headlights" look when asked some questions that he doesn't have an immediate answer for. Often, out of sympathy, the interviewer will say, "OK, don't worry about it, we'll move on to the next question." The best thing to do is persist to have him either answer the question or pass on it, but don't let him slide too easily. When the candidate is getting flustered or upset, calmly tell him to take his time and try to come up with an answer, or if he has no idea, he can pass. Be wary of the candidate that can never admit he doesn't know something and tries to "fake" his way through a topic. However, remember that the stress interview technique is not well suited for software developers. Give the super-nervous candidate a little reassurance and try to get him to calm down to complete the interview. After all, you don't want to eliminate a potentially good candidate due to some interview jitters. Of course, if one of your critical attributes is handling some pressure situations, a candidate who flusters when you are not even trying to stress him may not be your ideal choice for that role.

Judging the Candidate's Responses

In various Olympic sports such as diving, a panel of judges observes a competitor perform and then renders a score. While the scoring is somewhat subjective, there are guidelines that the judges follow to help maintain fairness and consistency during the competition. For example, in a dive if the toes aren't pointed during the entry into the water, the judges may deduct points. Exactly how many points they deduct may be subjective, but the important concept is that there are established criteria, things the judges are looking for that help them determine the scoring.

As an interviewer, you have similar goals as the judges. You want to maintain a consistent, fair judging mechanism that can be applied by all interviewers to all candidates. In some of the technical questions, it is very clear if the candidate answered correctly. With other questions it will be more subjective. Nevertheless, it is worthwhile to get a system in place for judging questions.

Table 6-3 provides an example system for judging responses to behavioral questions. Based on the questions you are going to ask, and the answers you are looking for, you can customize this to fit your needs. This example assumes each candidate starts with 10 points, and then points are deducted or awarded for various reasons.

Table 6-3

CANDIDATE'S RESPONSE	POINTS
Candidate clearly and concisely explained the situation, the action he took, and the result. The action was proactive and insightful, and the result was successful. Essentially a perfect answer.	Award 2 points
Candidate wasn't clear in describing the situation, action, or result.	Deduct 1 point
The action that the candidate took was not very good, was not proactive, or was not in the interest of the department or company goals.	Deduct 2 points

The candidate did not appear to learn anything from the situation.	Deduct 2 points
The candidate placed blame on others and did not accept responsibility for some things that were in his control.	Deduct 3 points

The main point is that you want to think about the judging criteria *before* you start interviewing. You also want to make sure all the interviewers on the panel understand the criteria so that there is some consistency in the evaluations among interviewers. Of course, since interviewing is complex and subjective by nature, there will still be some variance among interviewers. However, by having a judging system in place, you help ensure fairness to all candidates and provide a common reference for comparing two candidates.

Summary

Keeping the interview styles and questioning techniques in mind helps you have a more effective interview. Here are a few tips to remember:

- Use a mixture of Behavioral and Situational questions to get good interview coverage of past experience and a sense of future performance

- Ask open questions where appropriate

- Remember the 80/20 rule. Let the candidate talk!

- Establish judging criteria beforehand

By this point, we have defined our "must have" qualities our candidate needs for the position, so we now know what we are looking for. We have prescreened our vast collection of resumes, and through reading the resumes and phone screening, we have done at least two "elimination rounds" and now have a shorter list of candidates that we would like to speak to in person. It is time to start planning the interview.

7   INTERVIEW PLANNING

The introduction to this book mentions that some consideration must be devoted to "selling the job." That is, just as you are evaluating the candidate, the candidate is also evaluating the company. In order to attract the best people, you need to make a good impression. One of the best ways of doing this is to conduct a professional interview.

In a poorly planned interview, there is chaos. Some of the people the candidate is supposed to meet with are not available. There may be constant interruptions during the interview, or it is held in a noisy setting. In order to maintain a professional appearance, these mistakes of interviewer etiquette must be avoided at all costs. By following a systematic approach, you can take steps to ensure a smooth flowing interview.

Before the Interview Starts

Preferably a day or so before the interview, it is critical that each interviewer thoroughly review the resume. Nothing is worse for the candidate or the interviewer than to have the interviewer scanning the resume frantically to find something to talk about during the interview. This makes a bad impression and wastes valuable time.

To remind the interviewer to bring them up for discussion during the interview, particular points of interest should be highlighted on the resume beforehand. At least a day or two prior to the interview, one person, typically the hiring manager, should coordinate the formation of the interview teams and each team's assignments. Given the assignments that they have, the interviewers should think of some questions to ask.

This might be in the form of an internal memo such as this:

Interview of Jimmy Buffoon: April 24.

Conference room 112

8:45 AM Welcome and introduction: Rudy Edwards

9:00 AM – 10:00 AM John Doe and Jane Smith

(Technical experience: Programming language questions, software concepts)

10:00 AM – 11:00 AM Paul Davis and Ted McNeil

(Domain Knowledge: Industry/Application knowledge, system design experience)

11:00 AM – 12:00 AM Carol Stanley and Vince Fox

(Professional Skills: communication skills, teamwork skills, soft skills evaluation)

12:00 PM – 1:00 PM Lunch (Rudy, Carol, and Paul)

1:00 PM – 1:30 PM Facility Tour (Paul)

1:30 PM – 2:00 PM wrap up, Rudy Edwards

Notice that the memo indicates the name of the candidate, the date and times, and the location. These things may seem simple, but each detail helps avoid potential confusion during the interview process.

As each interviewer begins his session with the candidate, he should start with a brief introduction. It is a nice touch for the interviewer to provide a business card so the candidate can keep track of all the people he is speaking to. Before starting, it is also good to check if the candidate needs a drink or restroom break.

Towards the end of your interview time, if the candidate seems like a good prospect, you will want to devote some time to selling the position. Tips for selling the job are covered in more detail later in this chapter.

The Welcome

As first impressions are often lasting impressions, welcoming the candidate is important. In a facility with a receptionist or a security booth, someone there should be notified of the candidate's arrival and who to page or call when the candidate arrives. An alternate name should be provided just in case the original person is not available.

The greeter is typically the hiring manager, human resources member, or one of the members of the department that the candidate is interviewing with. To make the candidate feel welcome and at ease, when the greeter arrives in the lobby to meet the candidate, he should greet the candidate in a friendly, professional manner. The candidate should be treated with the same courtesy and respect that is extended to a client. The greeter should ask if the candidate would like to use the restroom and offer a beverage such as water, soda, or coffee.

The greeter should then escort the candidate to the first stop. Typically, this is the hiring manager's office. The hiring manager should give a brief overview of the position, optionally providing the candidate with a written schedule of the interview.

An example of such a schedule is shown below. Of course, depending on your department, there may be more or fewer steps. These steps are now addressed in more detail.

8:45 AM Welcome and introduction: Rudy Edwards
9:00 AM – 10:00 AM John Doe and Jane Smith
10:00 AM – 11:00 AM Paul Davis and Ted McNeil
11:00 AM – 12:00 AM Carol Stanley and Vince Fox
12:00 PM – 1:00 PM Lunch
1:00 PM – 1:30 PM Facility Tour
1:30 PM – 2:00 PM wrap up, Rudy Edwards

One disadvantage of providing the schedule up front is that it makes it harder to "abort mission" if the candidate clearly isn't going to work out. When you have a whole day interview planned, sometimes it happens that you realize in the morning session that things aren't going to work out. Why waste people's time interviewing a candidate you know you won't hire? There are a few ways to handle this. You really don't want to say, "Well, I think we've seen enough of you, goodbye." That would be awkward.

An alternative way to handle this situation is to ask the candidate to be prepared for a full day interview. Mention, however, that you are trying to line up several interviewers, but it depends on their schedules and you aren't exactly sure how things will line up for the afternoon. This will give you an "exit strategy" should you need it.

Once the candidate has interviewed with the first few people, the manager can get feedback on if it is worth proceeding with the afternoon session. If the candidate clearly doesn't fit, you can gracefully end the interview prematurely. In the example above, this technique allows Jane Smith to tell Rudy Edwards "Hey, this Buffoon fellow is a chump. Get him out of here so we don't waste any more time." Mr. Buffoon is less likely to realize he is being exited early, avoiding the awkwardness of the "abort mission" scenario. Regardless of whether you chose to provide the schedule to the candidate, you should always make a schedule and provide it to all the interviewers so that they all know who is next in line to speak with the candidate.

In the example schedule above, Rudy Edwards is the hiring manager, and the other six people on the schedule are interviewers that will be evaluating the attributes of the candidate. Ideally, a majority of the people will be from the same department or group as the position that the candidate is interviewing for. Just to have a different perspective, it is sometimes desirable to have at least one person from another group to participate in the interview. Rudy greets the candidate and explains a bit about the company, department, and position. Since the candidate and Rudy will meet again at the end of the interview and have more time for questions, this should not take more than 15 minutes.

The other six people interview the candidate in pairs, forming three separate interview groups. By using six people, there is some leeway if one or two people are unavailable to participate in the interview. Each interview group will focus on a different set of critical attributes that are determined to be necessary for the position. At the end, the candidate goes to lunch, typically with the hiring manager and possibly one or two other interviewers. By doing a facility tour after most of the interviews have been completed, the candidate has a better understanding of what he is seeing on the tour. If possible, a tour of the work area, labs, cafeteria, and any other areas of interest should be shown. Of course, if some areas are restricted due to security reasons, that may not be possible. After the tour, the candidate is returned to the hiring manager for any wrap-up questions.

As mentioned earlier, each interview group focuses on a different set of critical attributes that are determined to be necessary for the position. These attributes are discussed in detail in previous chapters. However, the attributes can be divided into three key areas:

- Technical Knowledge (Knowledge of programming languages, software development principles)

- Domain Knowledge (Knowledge of the industry and applications)

- Professional Skills (Communication skills, Teamwork skills, etc....)

Note that there may be overlap in these areas, and any interviewer may ask a question in any area that seems appropriate during the course of the interview. In other words, given the example above, if Carol Stanley and Vince Fox are assigned to evaluate Domain Knowledge, it doesn't mean they are prohibited from asking a question that probes technical knowledge, should it come up during the interview. The assignments are simply guidelines to help ensure thorough interview coverage.

Using the two-person format for each group, it is preferable that while one interviewer delivers the questions, the other interviewer is focusing on the candidate's responses and taking notes. By having one interviewer tasked with observation, it ensures that subtleties of the candidate's responses, as well as non-verbal communication (e.g., body language, posture, tone of voice), are noted. Of course, the "observer" interviewer is also allowed to ask questions. However, the majority of questions are typically asked by one interviewer, allowing the other interviewer to perform note taking and observation. In case the other interviewer is out of the office that day, each interviewer should be prepared to conduct the interview alone. In this way it is possible to maintain the time slot and keep the schedule on track.

Selling the Job

A portion of the interview time should go to selling the job and company. After all, just as you are evaluating the candidate, the candidate is evaluating the position. Therefore, you should be prepared to give a short talk about the positive attributes of the position and company. Examples of positive attributes include:

- Career Growth Opportunities

- Chance to Learn New Technologies

- Pleasant/Flexible Working Environment

- Recent Performance of the Company

- Commitment to Employee Growth (e.g., Company Training, Tuition Reimbursement)

- Selling the Area (if the candidate is from out of town)

In order to make the best sales pitch, it is useful to know what has motivated the candidate to come and talk to you today. Before starting the sales pitch, it may be helpful to ask the candidate why he is looking for a job and what he is looking for in his next position. Use his answers to focus your sales pitch appropriately. However, be cautious not to over exaggerate the positives. If you tell the candidate that a flexible start time is allowed but the actual allowable start time range is 7:55AM – 8:03AM, the candidate may be disappointed when he comes to work at your company. Don't oversell the position. It is much better to be truthful and up front about the position information. The cost of replacing an employee has already been noted. The last thing you want to do is give misleading information so that a candidate joins your company, only to quit a short while later.

Seven Tips for Effective Interview Planning:

1) Just in case the main contact is unavailable when the candidate arrives, provide an alternate contact for the candidate.

2) Some time in advance of the day of the interview, make an interview schedule.

3) Determine which areas of candidate experience each interviewer will focus on.

4) Should you find the candidate to be a good prospect, be prepared to sell the position and the company.

5) Prior to the day of the interview, arrange an appropriate place to conduct the interview, such as a conference room.

6) Prior to the day of the interview, review the resume, highlighting possible discussion topics.

7) When your time period starts, make every effort to be available and ready to go.

Using these tips will help make your interview go smoothly. Now, we are ready to talk about the interview itself.

8 INTERVIEW QUESTIONS

This chapter presents some questions to help form your interview plan. Technology changes fast, and new programming languages, protocols, and hardware platforms are always emerging. Therefore, this chapter does not attempt to provide a comprehensive list of actual technical questions to use (they would be out of date quickly), but rather to illustrate the techniques of constructing effective technical questions. Of course, if some of the questions shown in this chapter are relevant to your needs, then it is fine to use them. In the examples provided, since many people are familiar with that language, the C language is used.

So that you may draft similar questions for the topics pertaining to your job requirements, the primary goal of this chapter is to illustrate the *types* of questions used. For example, if you are hiring a Java developer, you should be able to draft appropriate Java questions that are similar in format to the questions presented here.

The following ten areas of interview questions are covered:

1) Questions You Should *Not* Ask
2) Programming Language Skills
3) Debugging Skills
4) Software Repair Skills
5) Development Environment Skills
6) General Problem Solving Skills
7) Communication Skills
8) Ability to Learn and Take Action
9) Discretion

10) Questions for Entry Level Candidates

Questions You Should <u>Not</u> Ask

Because knowing this is just as important as knowing what you should ask, this chapter starts with questions you should *not* ask. In the United States and Canada employee discrimination is illegal. This section is not intended to be a legal reference, but merely to outline some of the more commonly asked illegal questions. If you have any questions about the appropriateness of a particular question, I strongly encourage you to ask your Human Resources department or legal department for assistance.

It is illegal to deny employment to someone on the basis of age, race, national origin, religion, or sex. Therefore, questions that probe these areas should be avoided.

Do not ask:

- about the candidate's nationality, birthplace, or ancestry
- where the candidate learned a foreign language
- about religious or political beliefs or affiliations
- what religious holidays the candidate celebrates
- about marital status
- about the candidate's children, or if he has children
- about the candidate's age

In some cases, the inappropriate questions are asked innocently and are not intentionally aimed at discrimination. There have been cases where someone would ask a candidate who obviously seemed like a non-native English speaker where he was from. Or asked a question along the lines of: "Hmm… Wigglesbah… that's an interesting surname, what kind of name is that?" These are questions that probe the candidate's ancestry and must be avoided.

In these cases, it could very well be that the interviewer was honestly just trying to make some conversation to break the ice. Nevertheless, conversation like that could open a company up to a lawsuit. While many managers and HR professionals receive training in this area, many software developers do not. Due to the legal implications, it is important that anyone who interviews a candidate have an awareness of questions that are off limits.

Note that there are proper ways to ask about the candidate's ability to meet the job requirements. For example, if the job requires periodic weekend work, do *NOT* ask, "Do you go to church on Sundays?"

However, you can you can state:

"This job requires periodic weekend work. Can you work weekends?"

The best way to keep your questions legal is to stick to the job requirements when asking the questions. The questions should be about the job requirements and not asking about a particular situation of the candidate. For example, if extensive travel is required, you can ask if the candidate can travel. Do not ask if he is married or has children.

Again, if you have any questions about the appropriateness of a particular question, you are strongly encouraged to ask your Human Resources department or legal department for assistance.

Common Questions and Custom Questions

Common questions are those that you give to every candidate, and custom questions are chosen for a particular candidate. Prior to the interview, it is recommended that you devise a set of common questions for the bulk of the questions. For questions that test software development knowledge, common questions should be used. By using common questions where possible, you help to ensure fairness of the interviewing process, and it makes it easier to compare one candidate to another when they are asked much of the same questions.

Custom questions are appropriate when you identify something on a candidate's resume that you want to ask about, such as a particular job experience or area of knowledge that the candidate has indicated on his resume. For example, if a candidate lists experience working with flight control systems on his resume, you may ask a question about that experience that you would not necessarily ask all candidates.

Programming Language Skills

Since writing code in a programming language is often one of the key duties of a software developer, it is no surprise that programming language skills are often evaluated. How important is proficiency in a specific language? That depends on your situation. In particular, if you need someone to start making fixes tomorrow, then you probably want high proficiency with a specific language. However, software developers are often capable of learning multiple programming or scripting languages. So don't necessarily disqualify a candidate because he is light on Java experience but has several years of C++ experience.

Often a candidate has various programming languages listed on his resume. However, does he really have experience with the language, or did he just read a book about it? The main goal here is to weed out the phonies, the people that read a book but didn't really use the language on a daily basis. In most of the examples that follow, C is used for the software questions. However, most of the concepts apply to other languages.

Keyword Test

One of the first questions to present during the programming language evaluation portion of the interview is the keyword test. In this test, a written list of keywords of the particular language is presented to the candidate. For example, in C here is a sample list of keywords:

```
for

while

switch

union

volatile

static

const

continue
```

It seems pretty basic; however, it's surprising how many people stumble on keywords. However, if they have programmed in all but trivial programs, they should have encountered most of these. The point of the keyword test is to eliminate phonies. The keyword list should consist mostly of everyday keywords that are used to get work done in that programming language. A few obscure keywords can be thrown in as well. However, remember that you are not trying to "stump the candidate." You are simply trying to establish a working knowledge of the programming language he claims to know (assuming he put it on his resume). If a candidate scores well on the keyword test, it doesn't necessarily mean he is a good candidate, but if he doesn't do well on the keyword test, it may be time to cut him loose. It depends on how he represents his level of the language. If he admits he studied it briefly or used it a long time ago, it may be acceptable. However, if he claims expertise yet is unfamiliar with many basic keywords, it raises questions about possible misrepresentation of skills.

The candidate should know what *most* of the keywords do. Regardless of language, if it is C++, Java, Perl, or any other language, a keyword test can be the first line of defense in screening out phonies. Because languages evolve rapidly, examples for various languages are not included. Since many programmers are familiar with that language, most examples are presented in C. You can then adapt the keyword test for your needs. For example, if you are testing C++ knowledge, you might put keywords such as *new* and *this* in the test. When making your own keyword test, ask yourself this question:

When using this language, what are the top ten keywords a programmer must be familiar with to write meaningful code?

Use this as your starting point. If you want to put in some advanced keywords to test advanced knowledge, that is not a bad idea. Just remember that the bulk of the keywords should be commonly used keywords.

The next thing that that should be tested is some basic programming constructs. These examples are in C. Most programming languages have these constructs, but obviously you can customize the basic programming constructs to the features of the language you are trying to test:

What is a fundamental difference between a while loop and a do-while loop?

I have found that most candidates know this, but if they don't, it's a red flag. You are looking for an answer that discusses where the test occurs and that the do-while loop will execute at least once.

A "written" test can then be given. Here is a sample problem:

Write a prototype for a function that takes as input 3 integers, and provides as output, total, and average.

A question such as this can reveal multiple things. It's amazing how many people write out the function when all that is asked for is the prototype. The sharp candidate will take the time to read the question and do what is asked. Since it is not intended as a means to analyze algorithm skills, the actual problem should be trivial. Therefore, there is no need to see the algorithm. You simply want to observe how he handles parameter passing and returning multiple values from a function. Something about a programmer's style can be determined by how he answers the question. Many will do something like

```
int total (int A, int B, int C, float *
avg);
```

Or possibly return both total and integer with a parameter. Others will make structures for passing the inputs and outputs. A few will use global variables to pass the data, which is not preferred. The point is, look at his solution and how it fits in with the way you would choose to do things. Some candidates haven't understood how parameters are passed or how to get two pieces of data from a function. That tells something about their true level of experience. So in summary, there are two key things that can be learned from this question:

1) Does he understand how to pass data in and out of functions?

2) Does he have a rough idea of the syntax of a function?

The code is not going to be typed into a compiler, and a few syntax mistakes are acceptable, but he should know roughly where the arguments go and how a return type is indicated.

Of course, however the candidate chooses to make the function, ask him to explain to you how the function is used. In other words, hopefully he has developed a handy "API" for you to use.

A final word about programming language skills: Most programmers that are proficient in more than one language can easily learn another, so don't over emphasize direct language skill. If a candidate has strong C++, Basic, and Pascal knowledge, but is light on Java, there is a reasonable chance he can learn Java quickly. Even if the programming language skills are light in the language that is used on your project, you need to consider your timeframe, the position, and the other attributes of the candidate to see if it makes sense to hire him. Just as you don't want to hire a dud, you don't want to falsely reject a good candidate just because he lacks language skills that he may be able to quickly pick up once on the job.

When designing your written "programming constructs" questions, ask yourself the following questions:

What types of constructs are commonly used in your environment? For example, linked lists, semaphores, interrupts, etc.... Ask questions based on those concepts that are most relevant to your job requirements.

Can the question be reasonably answered in the time allotted? Be wary of drafting a question with so many factors to consider that it is difficult to answer it in a reasonable amount of time.

Debugging Skills

If the candidate will be doing some maintenance programming, questions having logic and memory errors are especially valuable.

The example below is a contrived piece of code that has logic errors and memory trouble. Hopefully, nobody would ever do this in practice, but it serves as a good starting point for discussion.

```
#define MESSAGE_LEN 5
char * error_message(int code)
{
 char message[MESSAGE_LEN];
 int i;
 for (i=0; i<= MESSAGE_LEN; i++)
  {
      message[i] = '\0';
  }
 if (code == 0)
 {
   strcpy(message,"PASS");
 }
 else if (code == 1)
 {
   strcpy(message,"FAIL");
 }
```

```
else
{
   strcpy(message,"UNKNOWN ERROR!");
}
return message;
}
```

The above function is riddled with errors and incompetence: How many errors can the candidate find? Within a few minutes of inspection, he should identify some memory overruns such as with the loop initialization, and the `strcpy` of "UNKNOWN ERROR." Returning the locally declared message should also catch his attention as suspect. Deduct points if he can't find any of the memory infractions. You can make your own examples that represent situations the candidate may encounter on the job. Given the available time, keep the examples fairly simple for the purposes of the interview. Don't hand the candidate a thirty page printout asking him to find the 400 mistakes contained within it.

There are other questions that should also be asked that don't rely on source code. The idea is to make some questions that require thought but are reasonably answerable during an interview where the candidate is not sitting in front of a computer. The keyword test, function writing, and basic programming constructs are elimination questions. They are "slow pitches" designed to eliminate someone who can't "swing the bat" too well.

It is worthwhile to assess the debugging skills of the candidate. Debugging often requires creativity and perseverance to find the problem and make the fix. This is important in many positions, especially that of an Implementer, Integrator, or Maintainer. Here are some things you'll want to ask:

Tell me about a difficult problem you had to debug recently.

Problems that were intermittent, only occurred in the field, or without debugging enabled are especially revealing. In any case, take note of what the candidate considered to be a "difficult" problem. If his idea of "difficult" is not something you would consider difficult, it helps put his debugging level in perspective.

Other questions along this line:

Did you ever encounter any problems that were timing related?

If the answer is "yes," follow with:

How did you go about finding the cause?

Here is another question to probe for creative debugging techniques. This is especially relevant for embedded systems where commonly used tools such as an integrated debugger and display console may not be available.

Other than the debugger, what other techniques might you use when tracking down a problem?

There are some obvious answers here, such as "add print statements" or add logging or post-mortem information (e.g., writing data to a file before terminating).

The answer, of course, depends on the environment the candidate has used. However, it is a plus if he has used something besides an integrated debugger to debug an issue.

Another topic to ask about is "lockups." When a computer program asserts on a line number, it is straightforward to know where the program stopped (although knowing why it stopped there may not be so easy). However, what happens when a program seems to lock up, but doesn't give a core dump, or assert, then what would the candidate do?

Did you ever encounter a program lockup, where no core dump, GPF, or assert was generated? How did you go about tracking down the problem?

More important than the actual answer given is whether he understands the situation. Many experienced developers have encountered a situation like this at one time or another. There is no wrong answer to how he solved it, but if he has never encountered it, or doesn't understand the question, it should be considered a warning sign as to his true level of experience.

Possible responses from the candidate may include the following:

- Using a profiler to observe process loading to see if any process is "spinning."

- Breaking in with a debugger and examining the state of the threads.

- Adding custom print statements (or logging) to track semaphore usage or program execution.

- Studying the circumstances that resulted in the lockup (assuming it was repeatable).

Written Tests

A few words about written tests are in order. First, when making the test, try to make problems that can be realistically grasped in the time allotted for the testing portion of the interview. Don't make questions that are syntax-centric. They don't provide much information. For example, testing a candidate's knowledge of the order of arguments passed into the `strncmp` function does not provide much value, since this is the type of thing many programmers look up as they need to, and they don't always memorize every function prototype. The idea behind these written tests is to see if the candidate has the gist of the problem and can formulate a good starting point. It should generally not be expected to write an answer to a test question that is able to be compiled without errors on the first try.

As mentioned previously, it is recommended that a similar test be administered to all candidates that are interviewing for a given position. This provides for a fair interview and makes it feasible to compare responses from multiple candidates. One advantage of giving a written test is that it is harder for a smooth talker to fake his way through a programming language test. Either he knows the stuff, or he doesn't.

Finally, when administering the test, be polite. Rather than just shove the test in his face, ask if it is OK to give them a short written test. Make it clear that this test is a standard procedure that all applicants are asked to take and that making a mistake or not answering a question doesn't necessarily disqualify him. The candidate may already be a bit nervous. The purpose of the test is not to intimidate or insult him, but simply to objectively assess how fluent he is in the topics of interest.

In the design of a written test, consider having the following sections:

1) Identify keywords

2) Write a function or prototype

3) Examine code to find mistakes

4) Illustrate knowledge of software concepts

Software Repair Skills

In various roles, such as an Integrator or Maintainer, the following are a few important skills that the candidate must have:

1) Quick learning of code he didn't write.

Some people get to know their own code well, but get a "deer-in-the-headlights" look when they have to work on someone else's code. Here are some questions to probe skill in these areas:

Suppose you are faced with making changes in a code base that is new for you. Assume there is minimal documentation outside of the code itself. How do you get yourself up to speed quickly with unfamiliar code?

Different people have different learning techniques that work for them, so, as long as he *has* an answer, there really is no wrong answer. Reasonable answers include the following:

- Graphing the modules in a structure chart, dataflow diagram, or some other ad-hoc graphical representation of libraries, function calls, data passing, timing diagrams, etc.
- Stepping through code with a debugger
- Reading through the code and distilling it to its basic functions.

There is no wrong answer. However, it is a plus if he has some system that works for him to get up to speed quickly. If a blank stare results from this question, it may indicate he is not methodical and is the type to jump in without having a system for getting his bearings in the software. If you are looking for someone to help with maintaining an existing program, this should be a concern.

2) Ability to appreciate the scope of changes.

On one hand, in a post-alpha or maintenance stage of a product, it is not always appropriate to do a "rewrite" of an entire module just because there is a bug in it or because the module is not well written. On the other hand, it is desirous to have someone diligent and thorough in tracking the root cause. The other extreme of the "rewriter" is the seat-of-the pants programmer who stops working the problem as soon as the symptom disappears. The person who says, "I put in this print statement, and the problem didn't happen anymore, so that was the fix" is probably not the right person either.

Maintenance programming requires someone with good judgment about analyzing risk versus reward when it comes to making changes. He understands that, given the time constraints, sometimes the most elegant solution is not appropriate. On the other hand, he knows he needs to get the best understanding of the problem he can and make sure he is minimizing impact on the existing code with the fix. Ask:

How do you determine the risk of a change?

Possible good replies may include the following:

- Considering number of functions/modules affected
- Considering the code path(s) affected and how often those code paths execute
- Considering the ability to test the change fully or the amount of possible test coverage that can be completed in the time allotted.

If you've been involved in a few projects, you have come to learn that a "one line" change doesn't necessarily mean small risk. Some one-line changes have been known to cause disasters. The candidate should recognize that a small code change is not necessarily a low risk code change.

Development Environment Skills

In most cases, the candidate will be part of a development team. You need to know how well he fits into your team and his configuration management habits and etiquette (if he has any). It is essential that a team member is aware of his actions. A bad check-in to the version control system could leave an entire group of programmers idle, so it is hoped he takes check-in policy seriously.

The extent to which these questions are asked depends on the environment. For example, in a military application there is often strict policy about how a developer makes a code delivery. In other cases it will be less formal. In either case, with the candidate, you are looking to assess his attention to detail and consideration of fellow programmers.

What version control system(s) did you use?

Hopefully, he will answer that he has used one (if he hasn't ever used one, ask if he is familiar with the purpose of version control systems). Regardless of which version control system the candidate has used, many of the basic principles of version control apply.

How do you decide when it is OK to commit a file?

The answer to this often depends on the phase of the project and the general policies in place at the company. For example, some places allow programmers to commit files at their discretion during implementation. Then as the project nears completion, commits are more controlled, possibly subject to project leader approval. Alternatively, the developer may be forced to commit to a branch that is then merged to the main branch by a build engineer.

Ideally, the candidate's answer should convey some understanding of how the code delivery policy worked in his company.

Policy aside, another aspect of this question entails how much unit testing the candidate performed before he felt that his file was ready for committing. Did he attempt to test every code path of a function? Did he have to simulate any error conditions to do the unit testing? If his answer doesn't touch upon any of this, go for a follow-up question such as the following:

What level of testing of your code changes did you perform before you felt it was ready to be committed?

The answer should convey something beyond "a gut feeling" or "I just knew when it was ready" and indicate some procedures or level of inspection that is performed before committing the file.

The next two questions probe how the candidate works in a group development environment.

Have you ever been involved in a project where two or more people might be making changes to the same file independent of each other?

If "yes," follow with:

What steps did you take to make sure your changes didn't interfere with the changes of other team members?

In many cases, the version control system can play a role in this, to make sure one doesn't commit changes that overwrite another's changes. So he can discuss that as a possible valid answer. Another possible answer may include discussing changes with the other developer before committing them. In any case, the candidate must be sensitive to these issues. The candidate who commits files without regard to his teammates can be a productivity killer for the whole team.

If he worked in a team environment, these issues should have come up, and he should have some sort of answer. If he doesn't know or doesn't have much to say on the topic, then it could be a red flag that he truly doesn't have much development experience or doesn't give much consideration when working with others.

Embedded Software Development

Embedded software development has a few key characteristics that distinguish it from other kinds of programming. If you are trying to fill such a position, you may want to ask questions that touch upon these characteristics. These characteristics may include the following:

- Increased debugging difficulty due to lack of conventional debugging tools

- Increased complexity of software/firmware upgrading than in a PC-based application

- Increased need for careful memory management, since there is typically less system memory available than in a PC-based application

- User Interface challenges due to reduced means for providing user feedback (Consider a product with only a four-line alphanumeric display. It will not have the same UI capabilities as a PC-based application).

Note that there are a wide variety of embedded systems, and some systems may not have one or more of these characteristics. In any case, when hiring an embedded developer, you may want to probe some of these areas.

What factors do you consider when designing a software upgrade mechanism for an embedded device?

Possible answers may include the following:

- The means for receiving the update (e.g., via network, memory card, etc....)

- The integrity checking of the upgrade software (e.g., version checking, checksum verification, etc....)

- Robustness (e.g., what happens when the upgrade fails, if the power is interrupted during the upgrade, etc....)

- Recovery (e.g., can the user revert to the previous software if the upgrade fails or is unsatisfactory?)

Tell me about a situation where you had to devise a scheme to save memory in one of the systems you worked on?

Possible answers may include the following:

- Using data compression or hashing to represent important data

- Processing data in smaller chunks to reduce the peak amount of memory required during program execution

Please describe the system startup process in one of the systems you worked on.

This is a good question to ask someone who has been involved in "board bring-up." The candidate should be able to explain the process, which may include details of how the processor bootstrapped, key hardware components initialized, and how the various software levels (e.g., bios, drivers, operating system, and application) were initialized.

There is no single right or wrong answer to these questions. They are designed to probe the candidate's insight and experience on issues that frequently come up in embedded development. Of course, these questions can be used for other software positions too. After all, issues such as memory usage and robust software upgrading are prevalent throughout the software industry.

Soft Skills

General Problem Solving Skills

The previous section spent time evaluating knowledge of specific programming languages that are required for the job. While it is desirable that the candidate be intimately familiar with the language to minimize the learning curve, that isn't enough. It is also necessary to gauge his general problem solving abilities. To do this, ask one or two non-language specific questions. Some may be in the form of "puzzle" type or what you might find in the analytical section of the LSAT or GRE test. Others may be situational engineering questions. Here are a few reasons to ask at least one question of this type:

Some candidates are quite proficient in their programming languages, but can't solve problems. They are only capable of coding what is specified. This is fine if you have your complete design written in detailed pseudo code, but if you want any designing done, these people are incapable.

Conversely, there are some very sharp people who may not have in depth familiarity with the programming language in question. You may not want to discard such a candidate just because he doesn't know what *volatile* means. By giving some non-language specific questions, these qualities can be assessed. Candidates that have most recently been managers but want to get back to software development may be rusty in their programming languages. For that reason, it can be valuable to ask some General Problem Solving Skills questions that do not rely on a specific programming language or technology. The motivation for this type of question is that a candidate with good problem solving skills should not be rejected just because he is rusty in his programming language.

Things are different, however, if you are hiring a contractor. With a contractor, someone paid to come in, do a job, and get out, then it is usually not appropriate to be this patient. In the case of a "hired gun," he must be proficient in the programming language that he is billing for. When filling a temporary position, a question of this type should not be asked. However, for a permanent employee, what is wanted is someone who is sharp. In many cases, sharp people who join a company without direct analogous experience are six months later contributing more than others who had years of experience in the programming language in question.

Because these questions can be somewhat stressful, a chance to see how the candidate performs under some pressure, such as when a release must be gotten out quickly can be observed. Just to give an idea of the style of such a question, a sample General Problem Solving Skills question is included below.

Sample Question:

There are three books in a box. Each book is on a different topic (horseback riding, knife sharpening, and sushi making). Each book has a different number of pages (100, 200, or 300 pages). The cover of each book is a different color (green, yellow, and white).

If you randomly select a book from the box, it will be either green, about knife sharpening, or 100 pages. If you randomly select another book from the box, it will be either white, about sushi making, or 200 pages. The sushi book has more pages than the horseback riding book.

What is the page count, subject, and color, of each book?

Answer:

The horseback riding book is 100 pages and white.

The knife sharpening book is 200 pages and yellow.

The sushi making book is 300 pages and green.

Solving the puzzle successfully is not the most important thing. The key thing to take note of is how he reacts when presented with the problem and the approach he takes to solve it. Here are some things to look for:

Administering the test:

Taking a test like this is stressful enough during an interview. So it is best to hand him the paper with the question and then leave the room for three to five minutes to give him a chance to solve it.

Assessing the Candidate's Reaction:

How does he seem when the test is presented? Some candidates may get a little angry or insulted. Maybe they feel it is not directly relevant to programming. This is a warning sign. A candidate with a "this is beneath me" attitude is not a good one. Other candidates may get in such a panic that they can't make any progress at all. A certain amount of nervousness or surprise at the question is to be expected. However, complete meltdown or anger about being asked such a meaningless question is a red flag that could disqualify the candidate.

The candidate's approach:

If he can solve it, that's great. However, even if he doesn't solve it, if he can show an approach, he can still earn some points. Was he able to at least eliminate some of the possibilities? For example, if he could say, "Well, I know the sushi making book is 300 pages," then at least he started to make some progress towards a solution.

There is the story of a candidate who aced the programming language test and was subsequently hired. As it turned out, the candidate was a total bust and was gone after six months. He could code very well, but he couldn't get a grasp of any sort of complex logic, and he was hired as a senior level engineer with software design responsibilities. A logic question might have screened out that candidate, saving a ton of trouble.

As long as the candidate shows a little grace under pressure and gives the question a reasonable try, he should not lose any points, and if he solves it or simply shows his approach, he gets some extra credit. The question does give some insight into:

- How he works under pressure

- His logical reasoning skills

- His attitude towards being asked to solve a puzzle

An alternative or in addition to a puzzle is a variation bearing relevance to an engineering application, a "factors" question. In a factors question, the candidate is asked to list some of the factors he would consider in approaching a problem, but not actually presenting a problem for him to solve. Here is an example:

What factors do you consider when determining the priority of a task in an RTOS environment?

Possible answers may include the following:

- How often the task needs to run

- If the task runs periodically or asynchronously

- How long the task takes to complete its work

- The semantic priority of the work performed (i.e., how important the work the task is doing is to the overall function of the system).

Of course, there are other possible answers. Assuming he is claiming to have some RTOS experience, it is expected that the candidate is to have some answers in this regard. These answers can serve as a springboard for some follow-up questions. If the candidate is unable to come up with any factors (or only comes up with one), then it provides insight into how much he has actually worked in this area.

During the resume review process, highlight areas of the candidate's resume and then develop some "factors" questions beforehand.

Communication Skills

Ah, it really is true. When I was a freshman in college, my English teacher said that communication was important, regardless of your career. Many engineering and science students think those liberal studies courses such as English and history are a waste of time, a means to an end, a requirement to "get through" so one can start doing "real" engineering stuff.

The truth is, in most engineering environments, engineers do a great deal of communicating. They communicate with their peers, their supervisor, with other departments within their company, vendor companies, and customer companies. There is great value on effective communication, and most positions require a reasonable amount.

Asking the candidate to discuss a recent project he worked on is a great way to assess both communication skills as well as technical experience.

Please explain one of your recent projects. Provide a block diagram view of the entire system.

Once he has drawn the system and given an introduction about it, follow up with a question such as:

Show me the parts you worked on, and how they related to the entire system and/or process.

This can provide a lot of valuable information. It reveals the candidate in terms of:

- Public speaking and explanation skills

- His understanding of the big picture of the system he worked on

- The true role he had in the system.

It's amazing to ask this question of someone who claimed to have designed a highly complex system, only to learn that the only thing he did was a small ancillary part such as a data logger or something like that and has no idea how most of the system worked.

When he is finished with his explanation, how well do you understand what is going on? Remember, the better he can explain things, the quicker he will be able to explain problems and issues to you and other colleagues on the job, which translates in to solving problems quicker and completing projects sooner, rather than later (or never!). Therefore, it is imperative that the communication skills of all candidates be evaluated.

Ability to Learn and Take Action

Being proactive is a desirable quality in almost any position. The "it's not my job" attitude isn't helpful in a software development environment. This is a difficult quality to gauge, but ask a question along these lines:

Tell me about a work situation where you felt you went "above and beyond" to accomplish a task.

The idea is to see what the candidate considers to be "above and beyond." Of course, if he can't think of such a time, that is also good information to have. The response he provides could also be a springboard for some follow up questions. For example, if the situation were a response to a crisis, you can ask:

Why do you think the problem arose?

In hindsight, what could have been done differently to prevent this situation?

The above two questions help gauge how well the candidate understands the situation and if he applies "post-mortem analysis" to prevent similar situations in the future.

What did you learn from that?

A desirable quality of a candidate in any position is the ability and desire to continuously learn. A "masked part" candidate must be avoided. Just like a masked ROM part can never be reprogrammed, because he feels he already knows everything, a masked part candidate isn't prone to learning new things.

If the candidate's answer doesn't indicate that something has been learned, that is a cause for concern. It may indicate that you are dealing with a "masked part" candidate, and, unless other factors make him desirable (e.g., if he is a Specialist), it is generally best to avoid these candidates.

Discretion

Depending on the role, discretion can be an important attribute. This is especially true in External Vendor and External Customer roles. In particular, you want to find out if this candidate has a large amount of internal "truth serum." A candidate with a large amount of internal "truth serum" tends to blurt out company private information outside of the company. This can be a problem if the candidate will be spending time with people from other companies, whether at trade shows, performing field support, or integration efforts with third party vendors. Ask a question along these lines to probe if internal truth serum is present in high levels:

Have you ever had a situation where you had to discuss an issue with a third party but couldn't divulge all the details?

An alternate phrasing is:

When dealing with people outside the company, how do you handle dissemination of sensitive information?

In the response, you are looking for an answer that touches upon factors such as the following:

- Determining ahead of time what is allowed to be discussed and what is off limits

- Disclosing only what is necessary

- Thinking about the implication of saying something before saying it.

This skill is important when dealing with other companies in a project manager role. There is a story about a situation where a project manager, fairly new on the job, revealed company internal manpower allocation information to a key customer. In particular, the project manager told the customer that there was only one person working on the project and that person was out of the office for a while, so he told the customer that his issues would have to wait.

What the project manager said was true. However, the customer, while not being given a specific headcount or names of people, had been under the impression that there were multiple people working on this project. The customer was naturally upset, and much internal havoc was caused to that company, trying to assure the customer that it had the resources to handle the work load. In this case, the information about how resources were allocated within the company did not need to be revealed to outside parties, but it was, and that caused trouble. Had the project manager instead escalated the issue internally, there may have been opportunity to put someone else on the project to help resolve the immediate customer issues without getting the customer upset. That project manager had too high a level of internal truth serum.

Note that it's called it internal "truth serum" because some people tend to blindly dispense information that should be kept private. This is *not* to advocate dishonesty towards anyone, or infer that one should look for a candidate who is willing to lie for the company. It is to say that in certain technical roles there is a need for discretion when communicating with external parties. This is difficult to find out during an interview, but if it is important for the position, it only makes sense to investigate it during the interview.

Questions for Entry Level Candidates

For entry level candidates, there is not much work experience to discuss. Therefore, more questions about academic life are appropriate:

What course did you like the most/least?

What was it about that course that made you like/dislike it?

I look for some enthusiasm from the candidate about the course that he liked. If he liked it because he found it interesting and it piqued his curiosity, that's good. If he liked it because he found the person sitting next to him to be attractive, in the scope of a job interview that's not a good reason.

What was your senior design project about?

This question gives the candidate a chance to demonstrate communication skills and explaining skills.

These days, many college students are getting work experience through a co-op or internship program. In that case, some of the work related questions mentioned previously can also be given.

Summary

As a result of reviewing this section, you should make up your own set of questions appropriate for the position and candidate. For the programming language part of it, chose keywords and problems appropriate for the language you are using. Focus on the attributes that are most important and ask questions in that regard.

Normally, these questions are given by multiple interview groups. Below is a sample interview plan.

Table 8-1

GROUP	AREAS COVERED
Group 1	Programming Language Skills, Debugging Skills, Software Repair Skills
Group 2	Development Environment Skills, General Problem Solving Skills, Discretion

Group 3	Communication Skills, Ability to Learn and Take Action
Manager	Wrap-Up

Having a plan such as this beforehand ensures that the interview coverage is optimal and the interview groups are asking questions to probe a variety of areas. The last thing you want is to have groups 1, 2, and 3 all asking similar questions about programming language skills with nobody evaluating other equally important skills.

Now it's time to give your interviews! The next chapter deals with the post-interview analysis and candidate selection issues.

9 CANDIDATE SELECTION

After you have completed an interview, and the candidate has left the building, you should have a wrap-up meeting to discuss the candidate. It is best to have the meeting that day, while details of the interview are fresh in everybody's mind.

The hiring manager should be the moderator for the meeting. Typically, the interviewers go around the room, giving their opinion, and recommendation. Sometimes, popular opinion will bias an interviewer. That is, if an interviewer does not like a candidate, but everyone else who speaks before him does, then that interviewer may feel pressure to give a positive response.

One possible way to alleviate this situation is to have each of the interviewers hand the moderator a slip of paper with his name and opinion (yes/no/maybe) before the start of the wrap-up meeting. Then, during the meeting, the moderator reads the paper and asks the named person to explain his position. For example, the moderator might read the slip and say, "OK, Vince, you said you liked the candidate, please tell us what areas you focused on with your interview questions and why you felt positive about the candidate." The interviewer can then share his experience with the group. It is certainly acceptable for an interviewer to change his mind after hearing some feedback from the other interviewers. However, an interviewer should not feel pressured to suppress his impression simply because he has a minority opinion.

Giving Your Vote

Sometimes there is tremendous pressure to hire someone. This can be to fill a job opening before a certain date (e.g., end of the fiscal year), to coincide with the kick-off of a large project, or to replace someone who is leaving. This pressure may be transferred to the individual interviewer to give a positive endorsement. Even with this pressure, it is important not to give a "thumbs up" to a candidate simply because "we need somebody here now." Remember the costs (both tangible and intangible) of making a bad hiring choice. If you truly believe a candidate is not a good match, don't make him a good match due to peer pressure or pressure from the hiring manager. Of course, sometimes you may be overruled, but at least your best assessment of the candidate's ability has been brought to the surface for consideration. In most cases, this voting is not done in a purely democratic environment. Even if there is not unanimous approval among the interviewers, the hiring manager may make the final decision to hire the candidate. One factor may be who the candidate will be working with. If someone in a different department gives a "no" vote, it may have different weight than if someone who will be working directly with the candidate gives a "no" vote. However, there should be serious reservations about hiring a candidate if the peers of the candidate have concerns about hiring him.

Supporting Your Vote

Regardless of your vote, try to support it with tangible information, such as the way the candidate handled a question or something the candidate volunteered during conversation. Sometimes people just get a "gut feeling" that a candidate is good or no good for a particular situation. One of the main points of this book has been to go beyond "gut feelings" and into the realm of tangibility. That is not to say that gut feelings are wrong or that they shouldn't be given credence. Gut feelings may very well be accurate. The key is to try to identify the root cause of these feelings. Did that candidate say something that gave you a bad feeling? Did he have a voice tone that indicated a lack of enthusiasm? Did he seem sincere and honest when answering questions? If you have a "gut feeling," try your best to identify the cause before reporting your vote in the wrap-up meeting. However, even if you can not identify the source of the gut feeling, share it in the meeting. It is possible that others may have had the same feeling, and the ensuing discussion may help identify the origins of your feeling.

Getting the Most from the Wrap-Up Meeting

Once each interviewer has given feedback on the specific interview area on which he focused, a composite view of the overall candidate can be formed. A rating scale, such as 1 – 5 or A – F is recommended. As discussed previously, it is also desirable to have a judging system in place. The judging system provides guidelines on how to score a candidate based on his responses. The specific ranking system doesn't matter, as long as the assessment in the various areas can be quantified. A candidate evaluation chart is then made. Refer to Table 9-1 for an example.

Table 9-1

Candidate Name	Jane Wilson
Date of Interview	May 16, 20XX
Interview Area	Rank
Programming Language Skills	A
Debugging Skills	B
Software Repair Skills	B
Development Environment Skills	A
Communication Skills	A
Domain Knowledge	C
Discretion	A

Other Comments	Candidate has extensive public speaking experience. Doesn't have much experience in our industry yet. However, may have good potential in a project management role in the future.

In some cases, it is immediately obvious that you want to hire the person or that you don't. In other cases, you may want to wait until you evaluate other candidates before making a decision. In this case, you can set the candidate evaluation charts aside until you have completed your current round of interviews. The candidate evaluation charts of multiple candidates can then be compared to make a more systematic comparison.

Once everybody has shared his thoughts about the candidate, the moderator should once again go around the room and ask the interviewers their vote on the candidate (yes/no/maybe). As mentioned previously, it is acceptable to have an interviewer change his mind after hearing about something the candidate said during another group's interview session. By asking about the candidate again after the discussion, you can get a more accurate evaluation from the interview team.

You should now have a greater awareness of factors that make a successful interview and the preplanning that is involved. As mentioned previously, this book is intended to be a supplement to the existing body of work on hiring practices with specific emphasis on hiring software developers. Therefore, you are strongly encouraged to review other fine books on this topic, some of which are listed in the following bibliography. As a summary, we end with the following keys to successful interviewing:

Before the interview

- Know the role of the person that you are hiring

- Know the critical attributes of that role

- Know what areas each interviewer will cover

- Thoroughly review the resume and formulate questions

- Be prepared to sell the position and company to the candidate.

During the interview

- Make the candidate feel welcome

- Let the candidate talk (Don't ramble on about yourself)

- Don't telegraph your opinions or otherwise guide the candidate to a particular response

Happy Hiring!

10 APPENDIX: Designing Your Interview

Table of Questions

The table below lists the major questions that are discussed throughout the text along with comments to serve as a reference for building your own list of technical questions.

Table 10-1

QUESTION	COMMENTS
How does your position fit in with the overall goals of your department and company?	Good general question to gauge the candidate's awareness of the department and company goals.
Explain a difficult situation and how you handled it.	An open question that serves as a starting point for discussion about the candidate's work experience.
Explain a time when something didn't work out as planned.	A question to investigate how the candidate handled something unexpected and learned from mistakes.

What was the toughest problem you ever had to debug?	Good question for an implementer or maintainer.
How would you handle it if a big release were due tomorrow and you found a serious problem the night before?	Good question for a team lead, manager, integrator, and implementer.
What would you do if you were integrating a third party library, and you found a problem in it that was a showstopper for your project?	Good question for team lead or integrator.
What would you do if you submitted a development schedule and were told it was too long?	Good question for a team lead, integrator, or implementer.
Tell me about a time you fixed a difficult problem under pressure.	An open question that serves as a starting point for discussion about the candidate's work experience.
In hindsight, what could you have done to prevent [the crisis] from occurring in the first place?	A question to investigate how the candidate learned from mistakes.

What is a fundamental difference between a while loop and a do-while loop?	Basic software knowledge question.
Tell me about a difficult problem you had to debug recently.	Good question for an implementer or maintainer.
Did you ever encounter any problems that were timing related?	General software knowledge question.
Other than the debugger, what other techniques might you use when tracking down a problem?	General software knowledge question.
Suppose you are faced with making changes in a code base that is new for you. Assume there is minimal documentation outside of the code itself. How do you get yourself up to speed quickly with unfamiliar code?	Good question for a maintainer.
How do you determine the risk of a change?	General software knowledge question

What version control system(s) did you use?	General software knowledge question.
How do you decide when it is OK to commit a file?	Probes configuration management practices.
What level of testing of your code changes did you perform before you felt it was ready to be committed?	Probes quality practices.
Have you ever been involved in a project where two or more people might be making changes to the same file independent of each other? *If the answer is "yes," then ask:* What steps did you take to make sure your changes didn't interfere with the changes of other team members?	Probes team development methodology.

What factors do you consider when determining the priority of a task in an RTOS environment?	Use a "factors" approach to assess knowledge without giving a specific problem to solve. Replace "RTOS environment" with your topic of interest.
Please explain one of your recent projects. Give a block diagram view of the entire system. *Then follow up with:* Show the parts you worked on, and how they related to the entire system and/or process.	Probes communication skills.
Tell about a work situation where you felt you went "above and beyond" to accomplish a task.	Probes the proactive-ness of the candidate.
What did you learn from that?	General follow-up question when the candidate reveals an experience.

When dealing with people outside the company, how do you handle dissemination of sensitive information?	Probes the discretion level of the candidate.
What course did you like the most/least? *Follow up with:* What was it about that course that made you like/dislike it?	Good question for an entry level candidate.
What was your senior design project about?	Good question for an entry level candidate.
Did you ever encounter a program lockup where no core dump, GPF, or assert was generated? How did you go about tracking down the problem?	Good question for an implementer or maintainer (or anyone who will be doing extensive debugging).
Have you mentored any more junior members? If so, what was your approach?	A question to probe the mentoring level of leadership experience.

Why do you want to leave your current job?	Probes likes and dislikes, and may provide insight into the attitude of the candidate.
How would you respond if a customer called you and simply said, "My system isn't working"?	Good question for an in-house or field support engineer.
Describe your ideal job and why.	Good question for ascertaining what motivates the candidate, and evaluating if that is in alignment with the job description.
What factors do you consider when designing a software upgrade mechanism for an embedded device?	Good question for an embedded software developer.
Tell me about a situation where you had to devise a scheme to save memory in one of the systems you worked on.	Good question for an embedded software developer, but also applies to other positions.

Please describe the system startup process in one of the systems you worked on.	Good question for an embedded software developer, but also applies to other positions.

11 BIBLIOGRAPHY

Beatty, Richard H. *Interviewing and Selecting High Performers: Every Manager's Guide to Effective Interviewing Techniques.* New York: John Wiley & Sons, 1994.

Fry, Ron. *Ask the Right Questions, Hire the Best People.* Franklin Lakes, N.J.: Career Press, 1999.

Rothman, Johanna. *Hiring the Best Knowledge Workers, Techies, and Nerds: The Secrets of Hiring Technical People.* New York: Dorset House, 2004.

Yate, Martin. *Hiring the Best: A Manager's Guide to Effective Interviewing.* Holbrook, MA.: Adams Media, 1994.

ABOUT THE AUTHOR

Michael Kahn has worked in software development for nearly two decades. He has worked in a wide range of environments, including defense projects, factory automation, instrumentation, and consumer electronics. He has worked for American, European, and Japanese companies and has worked for both large companies having over 10,000 employees and small companies with less than 100. This wide range of applications and corporate cultures has provided the background on which much the information in this book was developed. He holds a B.S.E.E. degree from Drexel University, and an M.S.E.E. from New Jersey Institute of Technology.

For more information about the book, please visit:

www.softwarehiringhandbook.com

www.ingramcontent.com/pod-product-compliance
Lightning Source LLC
Chambersburg PA
CBHW051054050326
40690CB00006B/717